How to design and develop business systems

A PRACTICAL APPROACH TO ANALYSIS, DESIGN, AND IMPLEMENTATION

Steve Eckols

Mike Murach & Associates, Inc.

4222 West Alamos, Suite 101
Fresno, California 93711
(209) 275-3335

Author's acknowledgements

As general editor, Mike Murach turned my rough manuscript into a polished book. More important, however, Mike has consistently provided me and my co-workers with a stimulating and flexible work environment and has encouraged us to develop and use new, more productive system development methods. For that, he deserves a large measure of credit.

Among my co-workers, Doug Lowe has been a constant source of innovative ideas and always constructive and pertinent criticisms. All of the methods in this book were improved by Doug, and several were his ideas to begin with. Wayne Clary developed and perfected the idea of using a structure chart at the system level. Anne Prince read the entire manuscript and suggested improvements throughout.

Finally, two of the users with whom I've worked deserve special mention. Jerry Holden and Jackie Christensen have been especially patient and tolerant of me and my methods. Special thanks to you all.

Steve Eckols
August, 1983

Development Team

General editor:	Mike Murach
Copy editor:	Carrie Gwynne
Cover designer:	Michael Rogondino
Production director:	Steve Ehlers
Artist:	Ed Gallock

Library of Congress Catalog Card Number: 83-62380

ISBN: 0-911625-14-3

Contents

Publisher's preface

In the last ten years, several authors have asked me to publish their books on systems development. But when I reviewed their plans and manuscripts, I found nothing that hadn't already been published many times over. Bureaucratic procedures, analytical approaches, documentation techniques ... ideas that in total don't begin to make up a practical method for the analysis, design, and implementation of a business system.

But perhaps the existing books and proposed books on the subject simply reflect the state of the art. Does anyone have a method for developing business systems that is both efficient and effective? If they do, you couldn't prove it by my experiences. I've seen systems developed that turned out to be two and one-half years behind schedule. I've seen systems developed that ended up ten times over budget. I've heard war story after war story about systems projects that failed, one more unbelievable than the next. But I've heard few success stories ... very few.

When we started developing the software for our first computer system about five years ago, we began by using traditional development methods, the ones you read about in books on systems analysis and design. In fact, we designed most of the functions of our first system using traditional methods. But as a small businessman, I soon realized that these techniques would defeat us. The techniques were both inefficient and ineffective. And they were completely impractical when it came to developing a modern, interactive system for a small business.

During the next five years, then, we evolved a new method for developing business systems. No one part of this method is completely new. Yet taken together, I think the ideas make up a complete method for the development of business systems, a method that is both efficient and effective.

To highlight the efficiency and effectiveness of this method, let me tell two stories. The author mentions the first one in chapter 1, so I will be brief about it. In nine months, working alone, the author of this book designed and developed a system for a commodity brokerage firm that consists of 128,000 lines of COBOL code. And this is by no means a simple system. The resulting system is a complete operational and informational system for the company. It is completely interactive, and it is nearly self-instructional. The user's manual is only 21 pages long, because all other instructions are stored in the system ready for recall by the user whenever needed. When I talked with the owner of the company a month or two ago, he said that the work that used to take them two days to do, now takes them two hours. And for the first time, he says, their operational data is reconciled to their general ledger data.

The second story is about our own system. After three years with a Datapoint ARC system, we realized that it was not going to be adequate for our future processing requirements. So we decided to replace it with a Wang VS system. Rather than simply convert the existing system, though, we wanted to enhance it (1) by correcting all the design weaknesses that resulted from the ineffective techniques we used when we first designed the system; (2) by providing far more information than the existing system was capable of providing; and (3) by taking advantage of the more advanced features of the Wang system. By the time we completed the analysis phase of this project using the techniques of this book, we realized that the old system had so many problems that it wasn't worth trying to use any of the existing code. So we developed the new system from scratch.

When finished, the new system consisted of 175,000 lines of COBOL code. But a subset of this system consisting of

125,000 lines of code was ready in six calendar months from the time we started the project ... and only two people worked on it. After installation, one person stayed on the project for three more months for training, debugging, and the development and installation of subsystems that were designed but not yet implemented. Like the brokerage system, our system is completely interactive and nearly self-instructional. Although I expected some problems following the installation of the system because the file interrelationships are complex, few problems developed, and the bugs all but disappeared within one month.

As I see it, these are success stories of quite dramatic proportions. Particularly, when I read yet another horror story about a mainframe system project that cost over $100,000 but was scrapped because it was less efficient than the manual system it was supposed to replace. Or, when I hear a programmer brag about developing 100,000 lines of COBOL code *in six years*. Or, when I read that the average COBOL programmer produces less than 5,000 lines of tested COBOL code per year. Or, when I go to a minicomputer trade show and watch a software package for the insurance industry go into a beeping loop right after the salesman tells me that the software took ten man-years to produce. Or, when I talk with the local ComputerLand salesman and find out that the general ledger package for the IBM PC is sadly unreliable, even though thousands of packages like this have already been sold. Yes, I think our stories are success stories by any standard. And we have to give much of the credit for the success to the development method we used.

What this book does

As the title suggests, this book presents a method for analyzing, designing, and developing business systems. The method is efficient because it allows you to develop systems faster than you've ever developed them before. It is effective because it helps you deliver systems that do what the users expect them to do, and more.

If you want, you can think of the method presented in this book as a new method. More realistically, though, you can think of it as a combination and refinement of ideas taken from many sources. To analyze existing systems and model new systems, our method makes use of data flow diagrams as promoted by Yourdon Inc. But it uses these diagrams in a simplified, practical manner. To design new systems, the method makes use of system structure charts. These are adaptations of program structure charts as promoted in the literature of structured programming. As you will see, the structure charts have a special usefulness when it comes to designing interactive systems, and they completely replace traditional documents like system flowcharts. To design file structures and databases, the method borrows from the literature of database design. But again, it applies the concepts in a simplified, practical manner. And so it goes. Nothing entirely new, except perhaps the composite method ... a method that is both efficient and effective.

When I say this book presents a complete method for developing systems, I mean a method for analyzing old systems, designing new systems, and implementing new systems. In other words, the focus is on systems, new and old, not political considerations within the business environment. So this book assumes that the new system is cost justified, agreed to by the user, and so on. As a result, it doesn't show you how to do feasibility studies, cost/benefit analyses, or similar tasks. Not that these aren't important subjects. It's just that many books already do an adequate job of covering them. But no book that I know of does an adequate job of showing you how to design and develop a system once you know that you need one.

Similarly, this book assumes that you can do tasks like getting information from the user and presenting your findings to the user. It also assumes that you have enough hardware and software knowledge to design and implement a new system. Again, these subjects are presented in many other books. And the author's intent is to provide new information, not to rehash old information. So he focuses on the method for analyzing, designing, and implementing business systems in chapter 1 ... and he keeps this focus throughout the book.

By keeping this focus, there's no mistaking the author's intentions. First, he shows you how to analyze an existing system, but only to the point that it is useful in terms of designing the new system. Next, he shows you how to model the new system using your analysis of the old system as a starting point. Then, he shows you how to use the model of the new system as a basis for defining data requirements, organizing the new system's functions, and designing the database. And so on. One phase leads to another without any digressions for subjects that aren't directly related to the development process. In just eight phases, you've analyzed, designed, and implemented a system. Although you may feel that none of these phases is completely new to you, I think you'll feel that all are new in terms of application and interpretation.

Who this book is for

When I finished reading the manuscript for this book, I said to myself: Anyone who has anything to do with the development of business software should read this book. That includes information and data processing managers, systems and programming managers, systems analysts and designers, and programmers. The book is particularly useful for people who are developing interactive systems, but the method of the book can be applied to any kind of system: batch or interactive; mainframe, minicomputer, or microcomputer.

Perhaps the book will prove to be most useful to people developing software in small programming shops. In these shops, there tends to be less bureaucratic pressure than in large shops, so the developers are more likely to change their procedures if they are confident that the changes will lead to better systems and improved development productivity. In particular, I expect the method of this book to be widely accepted by people who are developing software for minicomputers and microcomputers. I'm thinking here of people developing software in-house, people developing software on a contract basis, and people developing packages for machines such as the Wang VS or the IBM PC. We've developed software in all three ways with outstanding results in terms of software reliability, ease of use, functional completeness, and development productivity.

In the large shop in the large company, of course, the wheels of progress tend to move more slowly, perhaps rightly so. As a result, I don't expect large shops to adopt the methods of this book in their entirety. But I hope that they will try them on a pilot project or two. Then, based on the results, they can incorporate some of the methods into their existing development procedures. And they can continue to adopt and refine the methods as they get more experience with them.

So there's no mistake about it, I'm not saying that the method in this book is a definitive procedure for developing systems. And the author certainly doesn't mean to imply this either. We too expect to refine our method as our experience with it increases, and I'm sure we will continue to do so for many years to come. All we're saying right now is that our method works just the way it is. We think it works better than any other method we've used or read about. And, at the least, it is a good starting point for improving your own development methods.

How to use this book

Perhaps the best way to read this book is as an idea book. For every phase of analysis, design, and implementation, this book presents some new ideas. If you like the ideas (and I guarantee you'll like some of them), you can incorporate them into your existing procedures for system development. If you do, you should experience some improvements in development productivity and system effectiveness.

On the other hand, if you like all of the ideas, you can read this book as a complete method for developing systems. If you do everything the book says, you should end up with better systems at a lower cost than you've ever experienced before. But still, you'll have to use our method within some larger framework.

For instance, the book presents an effective method for analyzing an existing system using data flow diagrams. But what do you do after you analyze it? The book uses the analysis data flow diagram as a basis for modelling the new system. But before you do this, you may want to present your analysis to management, decide to increase or reduce the scope of the project based on the analysis, or evaluate the costs of the old system. Since all of these tasks are beyond the scope of this book and its method, you can see that you must use our method in the context of a larger working environment and a larger procedure.

Why this book is effective

Developing a method is one thing; writing about it is another. But I think this is an effective book for three main reasons. First, the author has emphasized the practical side of system development. As a result, this is not a book of theory; it is a book of proven techniques.

Second, the author illustrates his techniques using examples from a system he actually developed. Specifically, he uses examples from the commodity brokerage system that I mentioned earlier. When I first read the introduction to the brokerage applications, I felt that they were too complicated for a book like this. But I soon realized that you can't present many of the critical aspects of a development method unless you show how to apply it to the complex interrelationships of a real system. In contrast, how many books on system development show their methods as applied to trivial development problems?

Third, the number of illustrations the author uses assures the effectiveness of this book. In total, there are more than 100 illustrations that show the application of the methods in this book to the brokerage system. When you're through reading the book, these illustrations themselves serve as a basis for review. To a large extent, they are self-explanatory.

Incidentally, we realize that the illustrations also make it difficult to read this book. Because there are so many, you have to do a considerable amount of page flipping and book turning as you move from text to illustration and back again. Unfortunately, there is no way to solve this problem in a single book. We thought about publishing a separate book of illustrations, but we ruled this out to keep your cost down. So if you think the idea of a separate illustration book is a good one, I'd appreciate hearing from you.

Conclusion

I'm proud to publish this book because I think it is an effective presentation of a systems development method that offers some hope for the future. I don't think this book is the last word on improving development methods...but it may be one of the first. And I think anyone who develops systems should read this book before he or she goes one more procedural step.

As always, I'm interested in your comments, questions, and criticisms ... about the method or about the book. If you have any, feel free to use the postage-paid comment form near the back of this book. And thanks.

Mike Murach
August, 1983

Part

An introduction to system development

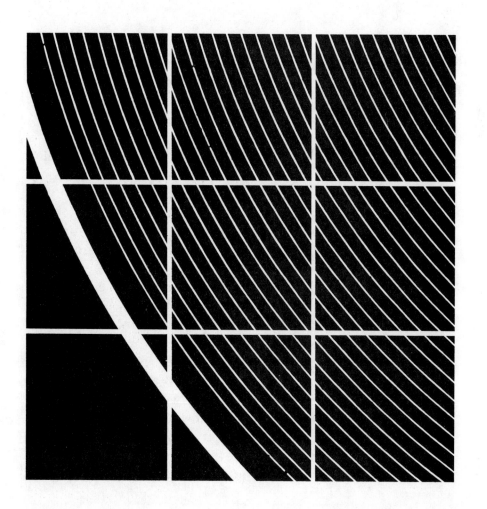

Chapter 1

A practical method
for developing business systems

 I hope we can agree that traditional methods for developing business systems leave a lot to be desired. Development projects are commonly "behind schedule and over budget." The users are often unhappy when the systems are finally installed because they're difficult to use or because they don't do what they were expected to do. It's not uncommon for systems to have serious "bugs" after they are installed so they require expensive modifications. In short, the methods used for developing business systems are all too often both inefficient and ineffective.

If you've worked on system development projects, you probably know what I'm talking about. With few exceptions, most companies need better methods for developing business systems. And they need them desperately.

In this book, then, I'm going to present a practical method for developing business systems. We developed this method during the last five years when we realized that traditional methods of system development were going to defeat us. Although we wanted to develop reliable systems that were easy to use, we also wanted to develop them at a rate of productivity that was unheard of at the time. As you might guess, we soon realized that we couldn't do either using traditional development methods.

During the last five years, though, we have used the methods presented in this book to develop systems for cotton and grain brokers, for retail distributors, for insurance administrators, and for our own company. In total, we have developed software consisting of more than 750,000 lines of COBOL code. All of the systems we've developed work with outstanding reliability. They're easy to use. They're inexpensive to maintain. And we developed them at rates of productivity that I hesitate to publish because they may seem unbelievable.

For instance, I developed the grain brokerage system that is used as an example throughout this book in nine months by myself. That includes analysis, design, and implementation. Since the system consists of about 128,000 lines of COBOL code packaged in 110 separate programs, that means I averaged more than 600 lines of tested COBOL code per day from the start of the project to its completion. And during the implementation phase of the project, I averaged more than 900 lines of tested code per day. After I installed the system at the end of nine months, I remained available to the user for another three months on a part-time basis for followthrough on training, debugging, and enhancements. But even adding three months to the development time, I still averaged more than 450 lines of tested code per day for the year of the project. If you compare my productivity on this project with the levels of productivity you're achieving in your shop, I think you'll agree that it's worth taking the time to improve your development methods.

The eight phases of our system development method

You can use this book in two ways. First, you can see it as a collection of techniques that you can adapt for use within your current procedures for system development. If, for example, you like the idea of using system structure charts but you don't think it's worth the effort to use data flow diagrams, you can use the one but not the other. In other words, you

---

Analysis

1. Create an analysis data flow diagram for the existing system.

2. Create a model data flow diagram for the new system.

Design

3. Define the data requirements.

4. Create a system structure chart.

5. Design the database.

6. Create the program specifications.

Implementation

7. Develop the programs of the system using structured programming techniques.

8. Document the system.

Figure 1-1   The eight phases of our system development method

can treat this as an idea book. If you do, I'm confident that you'll get some ideas that will help you improve your development procedures.

The other alternative is to accept our method in its entirety. As I have said, our method works. So if you do everything I recommend in this book, I believe you will improve the quality of your systems as well as your productivity.

In any event, figure 1-1 lists the eight phases of our development method. I call them *phases* rather than *steps* because you only do the first two phases in sequence. Once you enter the design and implementation stages of a project, you normally overlap the phases of development. I will explain this in more detail later on in the chapter.

Phase 1:   Create an analysis data flow diagram for the existing system   In recent years, system developers have come up with a simple yet effective tool for systems analysis: the *data flow diagram*, or *DFD*. The DFD is a drawing that shows what information comes into a system, what information goes out, and where that information is transformed in between. Although the DFD forces you to break a system

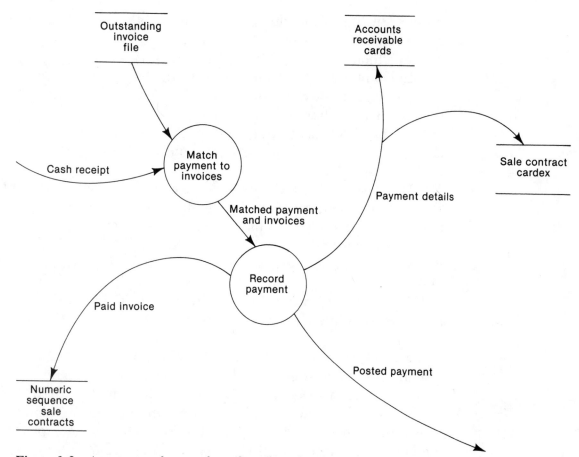

Figure 1-2   A segment of an analysis data flow diagram

down into manageable parts, it doesn't let you lose track of
the relationships between the parts.

An *analysis DFD* shows how an existing system functions.
Even if you've never been exposed to an analysis DFD, I
think you'll understand what one represents. For instance,
look at figure 1-2. This is a segment of a larger DFD that
describes the operation of the brokerage system example that I
use throughout this book.

As you can see by reading the names I assigned to the arrows, circles, and paired lines in this diagram, it represents
cash-receipts handling. Even though you aren't familiar with
the details of the application, I think you'll agree that the
diagram is easy to understand. What is missing from this example are the connections between it and the rest of the

system. But it's in showing those connections, which are so important in systems work, that the DFD is an especially valuable tool.

In chapter 2, I'll introduce you to the terminology and concepts of data flow diagrams and show you how to draw them. Then, in chapter 3, I'll show you how to create an analysis DFD for an existing system. By that time, you will clearly see how DFDs show the relationships between the parts of a system.

When you use a DFD, you can clarify, in a few strokes of a pen or pencil, concepts that might remain ambiguous using traditional analysis techniques. And the more complex the situation under study, the more valuable the DFD becomes. Perhaps most important, however, is that the DFD is meaningful to users. In my experience, most users welcome the use of DFDs because it's easy for them to recognize their own functions and data elements on the DFDs.

Phase 2:   Create a model data flow diagram for the new system   A *model DFD* shows how a new system will function. To create a model DFD, you start with an analysis DFD and reapply the same thought processes you used to create it. This time, however, you integrate the user requirements for the new system into the DFD.

Figure 1-3 is an example of a model DFD. As you will see, a model DFD for a new system becomes the starting point for identifying the new system's functions and data structures. As a result, the model DFD is critical to the success of the development effort.

In chapter 4, you'll learn how to create a model DFD for a new system. As you begin to model the system, it's essential that you emphasize key points and not let yourself be distracted by details. So I'll show you what nonessential details you can safely omit from the model DFD. Similarly, database design is a key part of system development. As a result, I'll show you how to draw the model DFD so you'll identify the collections of information that are central to the system. Later, these will become files or database views.

Phase 3:   Define the data requirements   In chapter 5, I'll show you how to keep track of the data requirements of a

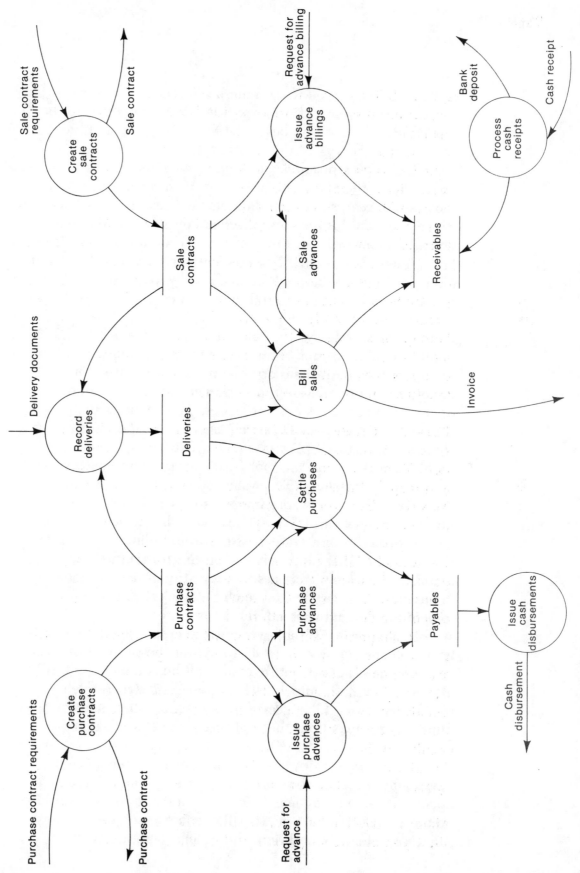

Figure 1-3 A model data flow diagram

system. To do so, you will create *preliminary contents lists* for important groupings of data you identify as you develop the model DFD. When you begin database design, you'll use these lists.

I'll also present a *data dictionary notation* that you can use to create these contents lists. This notation allows you to show relationships between data elements clearly and easily. These relationships are critical when you start database design. As you will see, the notation I recommend consists of just a few symbols and a few conventions for using them. But the resulting data dictionary entries will provide a solid basis for designing the files or database of your system.

Phase 4:   Create a system structure chart   The model DFD shows the relationships between the processes and data structures that are central and critical in a system. After you've established those relationships, you need to examine and organize all of the system's processes. This includes the processes on the model DFD as well as those you purposefully excluded from it to simplify it.

Figure 1-4 shows the tool I use to organize processes: the *system structure chart*. The system structure chart is a chart that shows all of a system's processes (figure 1-4 is a partial system structure chart). Note, however, that it doesn't show the relationships of those processes through data elements as the DFD does. Instead, it shows the control relationships among the system's processes by indicating levels of subordination. The system structure chart, like the DFD, is something users understand and appreciate.

In chapter 6, I'll show you how to create structure charts using a four-step procedure. For any but the simplest of systems, the system structure chart will be complex. So I'll also give you some guidelines for developing structure charts for large systems. I'll suggest formatting, naming, and numbering conventions that will make the charts easier to create and use.

The system structure chart is a critical document in our development method because it becomes the major planning document for the physical system. In most instances, each box on the system structure chart will become a program in the resulting system. In addition, the system structure chart is

Brokerage firm system structure chart
page 1 of 11

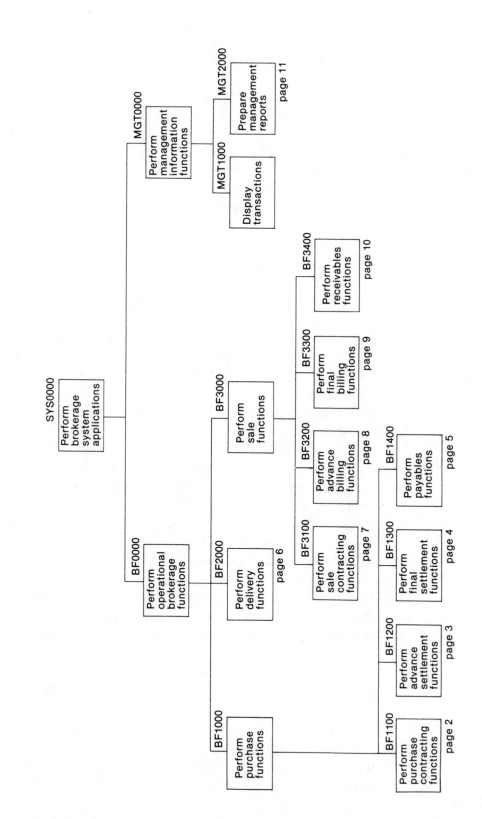

Figure 1-4   The first page of a system structure chart

valuable in planning the sequence of development of the system and for monitoring progress during the implementation of the system.

When you develop interactive systems, the system structure chart becomes the best design tool I've ever seen for developing a sensible menu structure. Since the structure chart shows the hierarchy of the program functions that make up a system, it can correspond to the accesses implemented in a menu-driven system. For instance, program BF0000 in figure 1-4 will probably be a menu program with four selections, one for each of the three programs it controls (BF1000, BF2000, and BF3000) and one to return to the previous menu. As I'll explain in chapter 6, a well planned menu structure can make a system easier to use, and it can be the basis for an effective security system.

Phase 5:   Design the database   In chapter 7, I'll give you some guidelines for developing the final plan for your system's files or database. Because database design is a technical undertaking that depends to some extent on the capabilities of your computer, I can't show you how to design a database in detail. But I can give you some useful general suggestions.

As you will see in chapter 7, the model DFD identifies most of the key operational files of your new system. However, you will need to identify collections of data you excluded from the model DFD to simplify it. After you've identified the files or database views your system will require, you'll need to determine how they'll be keyed. In other words, you'll need to determine the access paths to them.

In chapter 7, I'll describe a method that will help you identify those paths and organize them using a *data access diagram* like the one in figure 1-5. A data access diagram shows the relationships between access paths and data structures.

I'll also show you how to develop *data hierarchy diagrams* like the one in figure 1-6. These diagrams will help you understand your system's data organization more clearly. As its name implies, a data hierarchy diagram shows the hierarchical relationships among data structures. By recognizing these relationships, you will have a better understanding of how to implement the database.

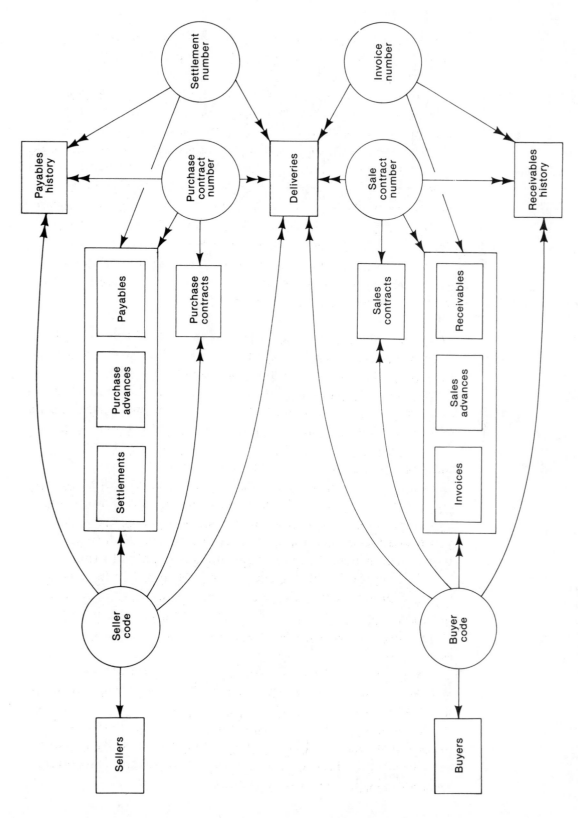

Figure 1-5 A data access diagram

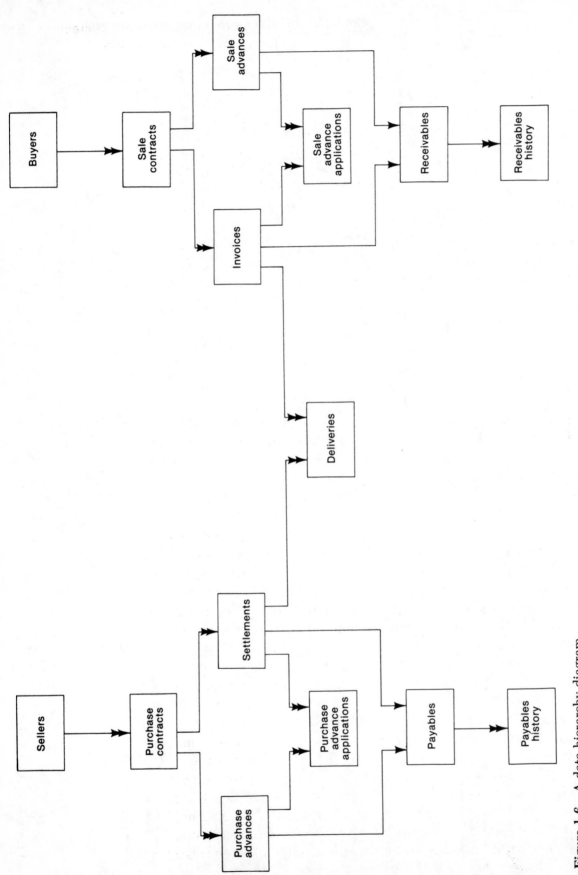

Figure 1-6 A data hierarchy diagram

After you've determined the physical organization of your system's database, your last step is to document that organization. You'll need to create record layouts and other supporting materials that will be available as the system is implemented. Fortunately, this task is mechanical. In chapter 7, then, I'll give you some helpful hints that will make that documentation more useful for implementing the system.

Incidentally, when I refer to a *database* in this book, I mean database in the general sense of the word. When implemented, it may be a collection of files in a small computer. Or, it may be a database in the specific sense of the word as implemented through database software. Either way, the techniques of chapters 5 and 7 will help you design the database.

Phase 6:   Create the program specifications   As you're deciding how the system's database will be organized, you will also begin to develop program specifications. It's from program specifications that you or your programming staff will implement the system you've designed. A complete set of program specifications includes a program overview and related report and screen layouts.

In chapter 8, I'll give you some guidelines for both report and screen design that stress consistency, readability, and ease of use. I'll particularly emphasize screen design because so many of today's systems are interactive. Since screen design can have a dramatic effect on a user's productivity and job satisfaction, my screen-design recommendations promote ease of use. I'll also show you how to plan and control the movement between screens when designing a multi-screen program.

After a program's screens and reports have been planned, the program's functions need to be identified. Through experience, I've found that the most useful and efficient program documentation is a program overview like the one in figure 1-7. A *program overview* is a listing of the input and output requirements of a program combined with the briefest possible, yet complete, description of what the program does. The description may include narrative, pseudocode, decision tables, even flowcharts. The overview ties the other items of documentation together and clears up specific questions about

Program: BF1110 CREATE PURCHASE CONTRACTS	**Page:** 1 of 3
Designer: Steve Eckols	**Date:** 6-15-83

Input/output specifications

File	Description	Use
PURCHCON	Purchase contract master file	Update
SELLERS	Seller reference file	Input
COMMODTY	Commodity reference file	Input
PCCNTRL	Purchase contract control file	Update
PCONTRCT	Print file: purchase contract copies	Output
PCREG	Print file: purchase contract register	Output

Process specifications

Get the system date using the subprogram SYSDATE

Get the system time using the subprogram SYSTIME

Do until operator indicates end of program by pressing PF16 from PURCHASE
    CONTRACT ENTRY SCREEN (in step 1):

1. Accept PURCHASE CONTRACT ENTRY SCREEN until no errors are detected in
      operator-entered fields or operator indicates end of program. Field editing rules
      are below. If the operator indicates end of program, terminate program
      immediately.

2. If the operator enters no price data on PURCHASE CONTRACT ENTRY SCREEN,
      accept PURCHASE CONTRACT BASIS PRICING SCREEN until no errors are
      detected in operator-entered fields. Field editing rules are below.

3. Format the PURCHASE CONTRACT VERIFICATION SCREEN:

    —Display the name and address of the seller specified by the operator.

    —Display the full description of the commodity specified by the operator (name, variety,
       and quality).

    —Format lines where the operator had several options from which to choose so they
       display only the one selected.

    —If the entry is for a basis contract, combine the pricing parameters selected from the
       PURCHASE CONTRACT BASIS PRICING SCREEN and format them as a
       sentence.

Figure 1-7   The first page of a program overview

program functions. It gives the programmer all the information he needs but allows some flexibility in implementation. In chapter 8, I'll show you how to prepare effective program overviews.

Phase 7: Develop the programs of the system using structured programming techniques After you've designed a system, the work of program development begins. Whether or not you will participate actively in it, you need to set some standards for this development. In our method, we recommend a detailed collection of program-development techniques that are generally referred to as *structured programming techniques*.

If you're not familiar with structured programming techniques, you ought to be. In a typical unstructured shop, programmer productivity is dismal, perhaps as low as 10 lines of tested code per programmer per day. Moreover, when a program is finally completed and put into production, it often requires on-going maintenance that can amount to more than its original development cost. In contrast, groups that use structured programming techniques have shown remarkable improvements in the quality and quantity of code produced by each programmer.

In chapter 9, I'll briefly explain and illustrate the concepts of structured program development, but I won't be able to present these techniques in detail. That, unfortunately, would require a book in itself. As you will see, structured programming is a complete methodology for program development that includes methods for design, coding, testing, and documentation.

When you use structured programming, one of the critical development documents is a *program structure chart* like the one shown in figure 1-8. In other words, one program structure chart is created for each program box on the system structure chart. The program chart shows the functional modules that the program is made up of, just as the system structure chart shows the programs that the system is made up of. When a program is coded, the boxes on the program structure chart identify the paragraphs in the program. As a result, by using the system structure chart in combination with the program structure charts, you have a complete directory to the program code of the system.

BF1110   Program structure chart
Page 1 of 2

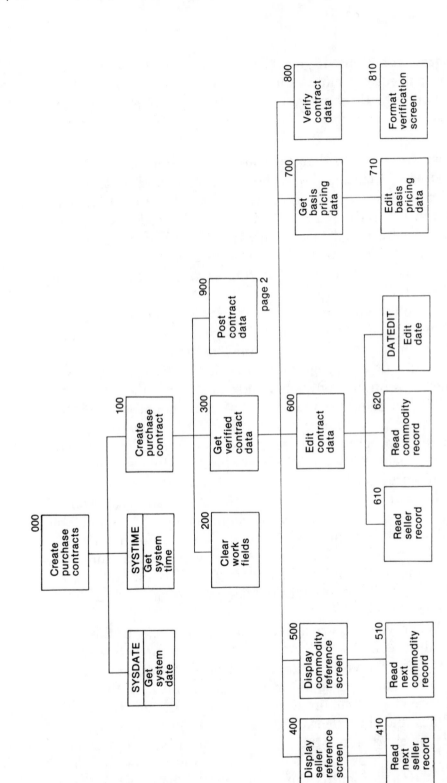

Figure 1-8 A program structure chart

Phase 8:   Document the system   Because businesses change, almost all systems require maintenance as they age. And that's making the liberal assumption that there are no bugs in the completed system. It's essential, then, that the persons doing the maintenance have access to complete, accurate, up-to-date documentation.

In our method, creating system documentation is a by-product of analysis and design. Although data flow diagrams and the system structure chart are critical for getting at the core of a system development problem, they are also valuable as documentation. Similarly, program overviews, program structure charts, and the program code itself are excellent documentation at the program level.

In chapter 10, I'll show you how to organize the system and program documentation that you create as you develop a system. I'll also recommend some additional pieces of documentation that you should create to complete the system documentation. What you'll end up with, I hope, is a set of documentation that is more useful, more complete, and less voluminous than any documentation you've ever created before. And I think you'll assemble this documentation in a fraction of the time you would take using other development methods.

Some related ideas

Do the eight phases in figure 1-1 seem simpler than the system-development phases you have been using? I hope so. Many of the development methods I have reviewed consist of bewildering sequences of steps and substeps and procedures within procedures. For instance, one book on system development presents an overall method that consists of 66 different steps. Another book uses 37 pages of data flow diagrams to present its method for developing systems. As I see it, though, a system development method just can't be that complicated. If it is, the likelihood of developing an effective system at a reasonable cost is slim indeed.

As you will see, the method presented in this book does suggest procedures for some of the development phases. For instance, in chapter 6, I'll show you a four-step procedure for developing a system structure chart. Even at that, though, I believe the method presented in this book is less complicated and more practical than the methods used for most development projects. But let me give you a little more perspective on our method before I continue.

Analysis, design, and implementation   As you can see in figure 1-1, you can divide the eight phases of our method into analysis, design, and implementation. Analysis consists of phases 1 and 2, and these phases correspond to what traditionally has been called "logical system design." When you complete the model data flow diagram for a new system, you still haven't made any decisions about how the system will be implemented on a specific computer system. That is, you haven't done the "physical system design."

As figure 1-1 indicates, I classify phases 3 through 6 as design and phases 7 and 8 as implementation. As you will see, however, it is possible and sometimes desirable to overlap design and implementation.

As part of implementation, you should realize that system installation isn't shown in figure 1-1 because it isn't covered in this book. I've omitted installation for several reasons. First, I have nothing to say about installation that isn't adequately covered in other books. And as much as possible, I want to present new information, not information that is available elsewhere. Second, I don't believe the installation of new systems is much of a problem in most businesses. In my experience, tasks like training operators to run systems and planning for conversion are done quite effectively in most businesses ... provided the systems work when they are installed. Third, if you adopt the methods recommended in this book, you will have fewer problems during installation than you have ever experienced before, so installation should run smoothly. Because you will be in contact with the users throughout the development process, the new system will do what the users expect it to do. Because the system will be implemented using structured techniques, it will be reliable. Because the system will be well documented, it will be easy to make whatever changes the users request during the first few months after installation.

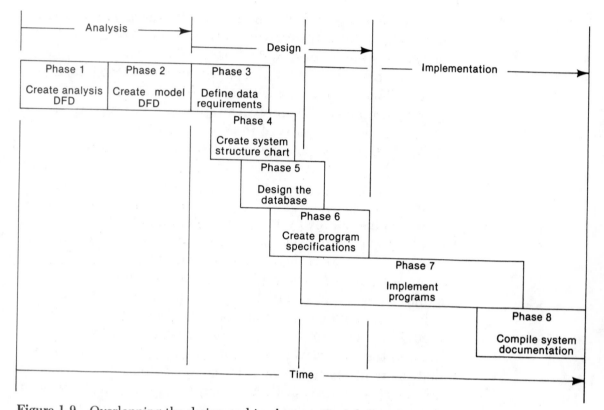

Figure 1-9   Overlapping the design and implementation phases of system development

Overlapping phases in the development process   As I said, the first two phases of our method are performed in sequence. That's because the output of the first phase (the analysis DFD) is input to the second phase. In contrast, the output of the second phase (the model DFD) is input to phases 3, 4, and 5. As a result, you can (and should) overlap phases 3 through 5 to some extent as you design and implement a new system. In addition, you can overlap phases 6 through 8 if you want to get an early start on system implementation.

Figure 1-9 schematically illustrates the overlap that is possible in stages 3 through 8. As you probably know, design tasks typically affect one another. For instance, you can't completely define a new system's database until you have a complete understanding of what all of its processes do. On the other hand, you can't completely specify the system's processes until you know how its database will be organized. As a result, you must work on database design and process specification together. As you do, you'll gradually approach a final design through successive refinement of both. Accordingly, their phases overlap.

Similarly, design phases can overlap with implementation phases. That simply means that if the design for part of a system is complete, its implementation can begin while design continues for other parts of the system. In fact, for a large project, that's how development probably should proceed.

How this method evolved   The method and procedures in this book grew from several years of study and experience. As we started to design the first system for our publishing company, we realized that traditional analysis and design techniques were inappropriate for our purposes. Because we had only 12 employees at the time, we realized that we didn't need a bureaucratic procedure for system development. Because the system was to be entirely interactive, we realized that many of the traditional development tools like the system flowchart were useless. Because we had limited manpower and budget, we realized that our productivity had to exceed typical productivity measures by a factor of ten times or more if our system was going to be cost-effective.

Fortunately, we were already familiar with structured programming methods because we had published two books by Paul Noll. Paul specializes in structured programming techniques for COBOL program development, and his techniques will definitely improve your programming productivity. At the start, then, we modified his techniques for our purposes, so program reliability and programming productivity were never problems for us.

On the other hand, we soon realized that system analysis and design can defeat your programming efforts. If you are shortsighted during analysis or design, it can mean many hours of reprogramming later on. As we were experiencing some problems like this, Wayne Clary, a member of our staff, suggested using a new document, the system structure chart, to represent all of the functions and programs required by a system. By using the system structure chart, we could make sure that we hadn't omitted any required functions because we could review the chart with the intended users. In addition, the structure chart served as a basis for estimating development times and for controlling the implementation of our system.

Although we were now able to develop functionally complete systems at a high rate of productivity, we still had some problems with file and database design. Not infrequently, we realized too late that the accesses to a database were inadequate or that required data elements were missing. And this in turn meant reprogramming. Although this type of problem wasn't as serious as a functional omission, we knew our method still had room for improvement.

At this time, I went to a Yourdon, Inc. class on analysis techniques, and I realized that the DFD provided a solid basis for analyzing data requirements as well as for simplifying system structures. As a result, we now use the DFD as the critical output document for the two analysis phases listed in figure 1-1. On the other hand, our use of DFDs is a much simplified version of the one recommended by Yourdon. As we see it, the entire Yourdon "structured analysis" methodology (of which data flow diagrams are only a part) is too burdensome for practical use. It requires too much unnecessary paperwork and compels the analyst to go over the same ground repeatedly and in excruciating detail. As a result, the use of DFDs to the extent recommended by Yourdon is likely to be as ineffective as traditional development methods.

For those of you who may want to consider using the Yourdon method in its entirety, I would like to offer one other comment about it. As I see it, a major problem with the "structured analysis" methodology is that it fails to bridge the gap between analysis and design effectively. Data flow diagrams are created at successively greater levels of detail until the process circles on the DFDs can be implemented as modules within the programs in the system. Unfortunately, this ignores many of the design considerations that are critical to the effective implementation of an interactive system.

In any event, once we began using DFDs as the basis for analysis and structure charts as the basis for design, the rest of our method evolved quite naturally. The data dictionary notation and database design diagrams in our method are based on similar tools used in the field of database administration. But we simplified them to make them more useful at the general design level.

In summary, the methods presented in this book are a combination of ideas taken from several sources. None of the ideas, however, has been left intact. Instead, we have modified the ideas to make a consistent, practical method that can be used for developing any kind of system: batch or interactive; inhouse or remote; mainframe, mini, or micro.

The benefits of using our method

I have already mentioned some of the benefits of using the method presented in this book. But let me be more specific. If you use our method, I'm confident that you will experience four main benefits in a measure you've never realized before.

User satisfaction through user involvement   Many times I've listened to users complain about the systems that have been presented to them. And I have to feel the main problem is that the user hasn't been involved throughout the development process. In traditional system development, it seems that the user is forgotten once the analysis portion of the job is done and the user has agreed upon a system specification. Then, six months or more later, the user is asked to review the completed system, at which time he realizes that the development group had an incomplete understanding of what he wanted.

In our development method, I recommend throughout that you review the development documents with the user. This is possible because the documents are developed with the user in mind. They use his language, they represent his current system, and they represent his future system. The DFDs and the system structure chart in particular are ideal for communicating with the user. In contrast, the traditional design documents tend to be in data processing rather than user language so they aren't useful for this purpose.

I also recommend communicating with the user during the implementation phases, and I recommend early implementation of critical portions of the system. Then, the user can review these portions of the system while it is still relatively inexpensive to make design changes. If you communicate with the user throughout the development process, there are no surprises when the system is finally installed. It does what the user expects because he has been involved in all stages of its development.

Increased development productivity   Although structured programming techniques can make dramatic improvements in programmer productivity, there's more to development productivity than this. By using a better analysis and design method, we want to shorten the development process by doing the analysis and design more efficiently. We also want to simplify programming and eliminate the need for reprogramming by doing the analysis and design more effectively. If we can design an efficient database and provide for all user functions before we start any program development, we can make dramatic improvements in our overall development productivity.

As I see it, the method presented in this book is efficient because it is a top-down development method. In other words, system development progresses by considering the most critical factors first, then the next most critical factors, and so on. Throughout the phases of development, I will advise you to defer trivial decisions until the more critical decisions have been made. This is possible because our method clearly proceeds from the top down. You start by identifying required user functions. Then, you organize these functions into a usable structure that identifies the programs within the structure. Finally, you develop the programs for these functions. In many cases, then, you can put off trivial details until you develop the program specifications for a function or until you are actually writing the program for a function. In contrast, many development methods try to gather all system information at once and then work with it. This is inefficient, and it leads to ineffective systems.

I think the methods used in this book lead to effective systems because of the development documents that are used. In particular, the DFDs, system structure charts, and data access diagrams are critical to our method. So whether or not you adopt our method in its entirety, I hope you will experiment with these development documents. In any environment, you can think of them as working papers and incorporate them into your current development procedures.

Early implementation of selected subsystems   When you finish the design phase for a system, you will often find that your development estimates far exceed the user's expectations. For instance, the user requests a system for installation in six months, but you realize that the system will require 18 months of work, and you only have two programmers you can assign to the project. So the best you can do is deliver a system in nine months, but that leaves no room for estimating errors. By using the methods presented here, though, you will often be able to identify subsystems that provide most of the benefits expected by the user but that take only a fraction of the total time to develop. Then, the subsystem can be delivered on time and the user will be satisfied.

This ability to select and implement usable subsystems depends to a large extent on the system structure chart. As you will see, it identifies all of the functions and programs of a system. Then, it is relatively easy to select functions from this chart that make a useful subsystem for the user. After the subsystem has been developed, the programming staff can complete the entire system and phase in new subsystems as they become available. In contrast, this type of phased implementation is often difficult to plan and control using traditional development methods.

Simplified system maintenance through improved documentation   As you will see in chapter 10, the documentation compiled when using our method is complete and understandable, but it isn't overwhelming. Perhaps the critical system document is the system structure chart because it is a directory for all of the programs and procedures in a system. The modules it represents point to program structure charts, and these charts in turn point to independent modules within the program code itself. So if some aspect of a system needs to be modified, it is easy to identify the programs that need to be changed as well as the modules within the programs that need to be changed.

In contrast to our complete and manageable system of documentation, traditional system documentation is often a nightmare. Several times in the past, I have been asked to review a system's documentation to see whether we would want to make enhancements to the software on a contract basis. On all of the occasions, I found that the program documentation wasn't too bad, but the system documentation was impossible to work with. If it existed at all, it was likely to be incomplete. So you could make changes to the programs without too much trouble, but it was almost impossible to determine which programs needed to be modified. If you work in a large company, you may be aware that more than one of its subsystems has been declared "unmaintainable" by the maintenance department because its system documentation is impossible to work with.

Summary

Whether you use some of the techniques recommended in this book as working papers or you adopt our method in its entirety, I'm confident this book will help you improve your system development efforts. Keep in mind, however, that it is not intended to be a definitive book on system development. The focus is on a development method, not on the skills and knowledge that are used within the method.

For instance, this book doesn't cover specific hardware configurations, although you need to understand hardware capabilities and limitations in order to design a system. Similarly, the book doesn't cover file-handling techniques, even though you need to know them in order to design an adequate database. As I hope you'll agree, these considerations vary too much from one shop to another to be included in this book. If I had included them, the book would be much larger and the emphasis would be taken away from the development method I'm recommending.

Similarly, I have omitted the more general (and less useful) topics that you find in college textbooks on analysis and design. For instance, I haven't covered communication skills or information gathering techniques. And I haven't covered cost/benefit analysis or decision table techniques. If you're interested in any of these topics or others I haven't covered, you can visit the library or the bookstore to review the available system development books. By avoiding these general topics, I have been able to focus on our development method.

In any event, I hope you now have a general idea of our method and approach. As I proceed through the eight phases of development, I try to focus on essential, critical points and defer non-critical details as long as I can. Usually, that means well into the design phases. In addition, I try to involve the user in the project as deeply and thoroughly as I can. Then, I'm confident that I'm always on the right track (or as close to the right track as possible).

Terminology

data flow diagram
DFD
analysis DFD
model DFD
preliminary contents list
data dictionary notation
system structure chart
data access diagram
data hierarchy diagram
program overview
structured programming techniques
program structure chart

Objectives

1. List the eight phases of our system development method.

2. List the four main benefits of our system development method.

Part

Analysis

Chapter 2

An introduction to data flow diagrams

 In this chapter, I'll present the background you need to use one of the most valuable development documents, the data flow diagram. After I present the parts of the data flow diagram and explain how one's created, I'll present some considerations you should keep in mind when you use data flow diagrams in your own projects. When you finish this chapter, you'll understand the use of data flow diagrams for the analysis of existing systems.

The parts of a data flow diagram

As I said in chapter 1, the *data flow diagram*, or *DFD*, is a drawing that shows what information comes into a system, what information goes out, and where that information is transformed within the system. Figure 2-1 is a simple DFD that represents a function common to almost all businesses: collecting money from customers and depositing that money in a bank account. You can recognize that just by looking at the diagram and reading the names I assigned to its parts. That's one of the advantages of the DFD: it directly suggests

Figure 2-1   A data flow diagram for processing customer payments

what might otherwise require many words of narrative to describe. Let me now describe the parts that make up a DFD.

Terminators   Every item of information that flows into a system must come from a *terminator*, and every item of information that flows out must end up at a terminator. By definition, then, a terminator is outside a system. A single terminator, represented by a rectangle, may be both a source and a destination of items of information. In figure 2-1, the two terminators are CUSTOMER and BANK.

Processes   In figure 2-1, the circle labelled PROCESS PAYMENTS represents a function, or *process*. A process, represented by a circle, is a place on a data flow diagram where an item of information is transformed in some way. In this case, PAYMENTs are transformed into a BANK DEPOSIT by the process called PROCESS PAYMENTS.

Data flows   To show the interrelationships that are so critical in analysis, the *data flow* is used. A data flow is a conceptual representation of an information link. To indicate a data flow, draw a line with an arrowhead showing the direction of the "flow." In figure 2-1, the data flows are PAYMENT and BANK DEPOSIT.

A data flow is always considered to be a single, discrete unit of information, but that unit may be complex, consisting of many subunits. And, as this example illustrates, a component of a data flow need not strictly be "data" in the data-processing sense. PAYMENT implies a check, something tangible, more than just "information." Similarly, BANK DEPOSIT is a collection of checks and a bank deposit slip.

Data stores   If I explore in greater depth what happens inside PROCESS PAYMENTS, I can expand my first diagram into the diagram in figure 2-2. This DFD shows in greater detail just what PROCESS PAYMENTS represents and uses the fourth and last of the parts of a DFD, the *data store*.

You can think of a data store in physical terms as a file (manual or automated) or a database view. I represent a data store at the end of a data flow by drawing two parallel lines. Figure 2-2 shows three data stores. Typically, a data store represents a holding area for information that's required by more than one process.

How to create a set of DFDs for a system

In its present state, the DFD in figure 2-1 is not particularly useful. For it to be of practical value, it's necessary to develop it in more detail. The way to develop a set of DFDs that describe a system is to *expand the processes* you initially identify. As you expand the processes, you must have a consistent method for naming the parts of the DFDs, numbering the processes, and making sure that you don't misrepresent the system. So let me now explain how to develop a set of DFDs for a system.

Levelling   Figure 2-2, which you just looked at, is equivalent to figure 2-1. Both represent handling payments from customers and creating a bank deposit. Can you understand what figure 2-2 represents without reading a detailed explanation of it? I hope so, because the diagram and the names assigned to its parts should be self-explanatory. The only difference between figures 2-1 and 2-2 is that the second diagram provides more detail than the first one does.

When you expand a process in this way to produce an equivalent, more detailed, lower-level diagram, it's called *levelling*. Here, the high-level process PROCESS PAYMENTS is refined to a lower-level diagram that contains three subordinate processes and three data stores. In turn, each of the three subordinate processes can be refined to an even lower level. Levelling can continue downward many levels. As a practical guideline, though, levelling to two or three levels

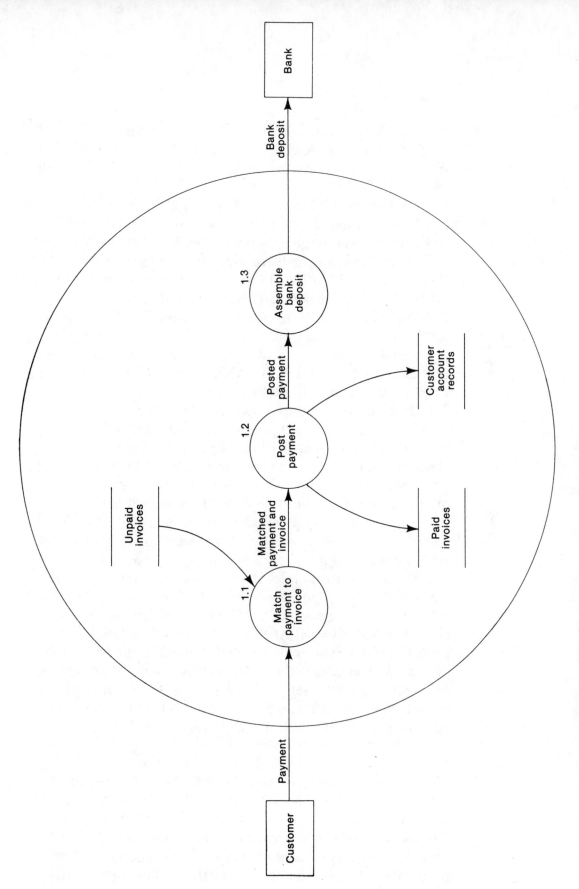

Figure 2-2 The expansion of the DFD for processing customer payments

deep is adequate for most purposes. Four or five levels starts
to get impractical. And six or more levels is likely to be both
unnecessary and overwhelming.

Conserving the data flows   When you expand a process using
levelling, its input and output data flows must be *conserved*.
For instance, the large circle in figure 2-2 containing the
three processes and the three data stores is the same as
PROCESS PAYMENTS in figure 2-1. One can replace the
other. To test their equivalence, you must make sure that
they have the same input and output data flows. Although
the large circle in figure 2-2 contains its own internal processes, data flows, and data stores, the data flows that cross
that circle (PAYMENT and BANK DEPOSIT) are the same as
those that enter and leave PROCESS PAYMENTS in figure
2-1. That means the data flows are conserved.

It's this simple concept that makes the DFD such an
effective analysis tool. If you make sure data flows are conserved as you go from one level to another when you expand
processes, you'll keep track of the vital interrelationships that
otherwise are so easy to lose sight of. Now that isn't to say
that you won't develop complex DFDs. You will. But the
complexity of each process won't be overwhelming, and its
relationships to other processes will remain obvious.

Suppose, for example, that you start with a high-level
DFD that has five processes, expand each of those into five
subordinate processes, and continue the expansion process
down two more levels. At that point, you'll have 625
processes. But if you make sure that the data flows for each
process are conserved, the relationships among those 625 processes will be fully described. Even for a significantly simpler
system, that would be no small accomplishment, something
you just couldn't do with narrative text.

Numbering the processes in a set of DFDs   If you expand a
DFD to one or more levels, you need to keep track of the processes within processes. Fortunately, this is relatively easy if
you use a clear system for numbering the processes.

Here's a simple numbering system you can use. First,
assign a number to each process of the top-level DFD. In

figure 2-1, the top-level DFD consists of only one process, so, of course, it's number 1. Then, label each of the *child processes* with a number such as 1.2 (for the second child process within process 1). You can see in figure 2-2 that that's how I numbered each of the three child processes I identified (1.1, 1.2, and 1.3). You can extend this numbering scheme down as many levels as necessary. For example, if process 1.2 were divided into three lower-level processes, they would be numbered 1.2.1, 1.2.2, and 1.2.3.

As you can imagine, this isn't an ideal numbering system. Perhaps none is. If you have to develop a DFD that's several levels deep, this numbering can get cumbersome. In general, though, you shouldn't have to go more than two or three levels deep. So this numbering system is effective most of the time.

Naming the parts of a DFD   When you create a DFD, you should try to give a meaningful name to each of its components. To a large extent, analysis is finding out all you need to know about a system's current or proposed functions and data. So it's important that you be able to identify those functions and data by name. If you find you can't give a meaningful name to a DFD component, that's a good indication that you don't fully understand that part of the system.

Be practical, though. As long as you understand what each DFD component represents, you can let the context of the diagram help clarify a name. Yes, you should try to assign an unambiguous name to every component in a DFD. But don't overdo it. At this point in the system development process, trying to assign the perfect name to every component can lead to a considerable amount of wasted energy.

In other books on analysis and design, you're likely to be warned *not* to use words like *process* and *data* in your names. The argument is that such words aren't descriptive enough. From a practical point of view, though, these words are acceptable at this stage in design provided *you* know what they mean. If, for example, you know that PROCESS PAYMENTS means match a check to an invoice, record the payment, and prepare a bank deposit, what difference does it make if *process* is a vague word? As I see it, little or none.

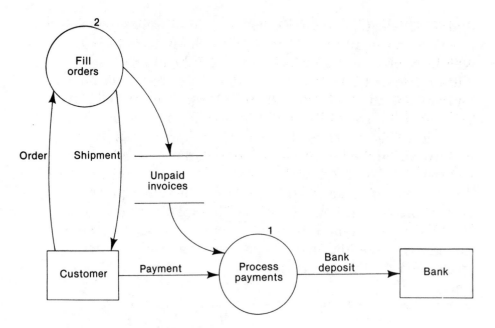

Figure 2-3 Using a data store to connect processes

Using data stores to connect processes A data store that has only data flows coming from it (and not going into it) is a danger sign when you're creating a set of DFDs. If information is to be extracted from a data store, it must be put there. If you look again at figure 2-2, you'll see that that's the case with the data store UNPAID INVOICES. The unpaid invoices that are matched with incoming payments have to come from somewhere. Since this diagram doesn't indicate where they come from, you can tell that it focuses on too small a segment of a larger system.

To identify where in the system unpaid invoices come from, it's necessary to look outside the narrow subsystem PROCESS PAYMENTS. Figure 2-3 presents the answer. Unpaid invoices are created by another process called FILL ORDERS that transforms its input (ORDER from the terminator CUSTOMER) into two outputs (SHIPMENT, which goes back to CUSTOMER, and an unpaid invoice that is placed in the data store UNPAID INVOICES).

To create the DFD in figure 2-3 from the DFD in figure 2-2, I removed the data store UNPAID INVOICES from PROCESS PAYMENTS to make it available not only to that process but also to FILL ORDERS. This illustrates the critical function of a data store: to act as an intermediate holding area for information shared by two or more processes. In other words, data stores are used to *connect processes*. Later, when you begin to consider database design, it's data stores that you'll look at with greatest interest.

You should notice in figure 2-3 that I didn't name the data flow that goes from FILL ORDERS to UNPAID INVOICES. That's because it's clearly identified in context by the names of the process and the data store it connects. Similarly, I didn't name the data flow from UNPAID INVOICES to PROCESS PAYMENTS.

Two general guidelines for developing DFDs

In general, you should use the methods of this book with two attitudes, or guidelines, in mind. First, you should continually involve the user in the development process. Second, you should put off dealing with trivial detail as long as possible in the development process. With that in mind, let me give you two guidelines for developing DFDs.

Develop the DFD with the user The most important advice I can offer for using DFDs is that you draw them with the user. When you do, don't hesitate to change them when you see the need to do so. After all, the DFD is a development tool that's intended to help you clarify your thoughts. So if you change a DFD based on a user's suggestions, it's just an indication that you're gaining a more complete understanding of the system.

When I say to involve the user, I mean to work directly with him. If you conduct a traditional interview, retreat to the peace of your desk, and then draw a DFD, you will certainly clarify your own thoughts. But though your thoughts may be clear, they may be wrong. So the time to develop a DFD is when you're in contact with the user. I've found through experience that you and your user working together will often think of important considerations that might not have come to mind otherwise.

Fortunately, DFDs are easy to redo. For example, figure 2-3 shows three data flows associated with PROCESS PAYMENTS rather than two as in figures 2-1 and 2-2. As a result, I redrew the DFD for PROCESS PAYMENTS in figure 2-2 to insure that the data flows are conserved. Figure 2-4 presents the revised DFD for PROCESS PAYMENTS with the large circle representing the new boundary of the process. It only took me a few minutes to redraw this DFD, and the improved view of the system justifies this effort.

So when you find an error in a DFD, change it right away. If necessary, don't hesitate to scrap a complete set of DFDs and create a new set. The experience gained in developing the first set won't be lost.

Put off the documentation of detail as long as you can
Whether a DFD component is simple or complex, at some point during the system development process, you have to understand each one. On the other hand, you can't let yourself get overwhelmed by detail. As much as possible, then, you must put off the documentation of detail until a later development stage. As a general rule, you should put off the documentation of detail as long as you possibly can.

I've already mentioned that you need to be able to name each DFD component but that you shouldn't waste time and energy in a compulsive effort to do a comprehensive job of it. Even more so, you shouldn't try now to document formally the detailed components of data flows, data stores, and processes. Instead, I suggest that you keep in mind that it's a necessary task but one that should be deferred as long as possible. If you're able to assign a meaningful name to each component of a DFD, you've done enough at this stage of development.

To some extent, however, your shop's standards will determine how long you can put off that kind of detailed documentation. If you work in a large shop, chances are you won't have much leeway in deciding how and when to document your analysis work. If you're in that situation, of course, you must do what the standards require.

But if you work in a less structured environment, you may be able to decide for yourself when to create detailed documentation. Now I'm not trying to convince you to

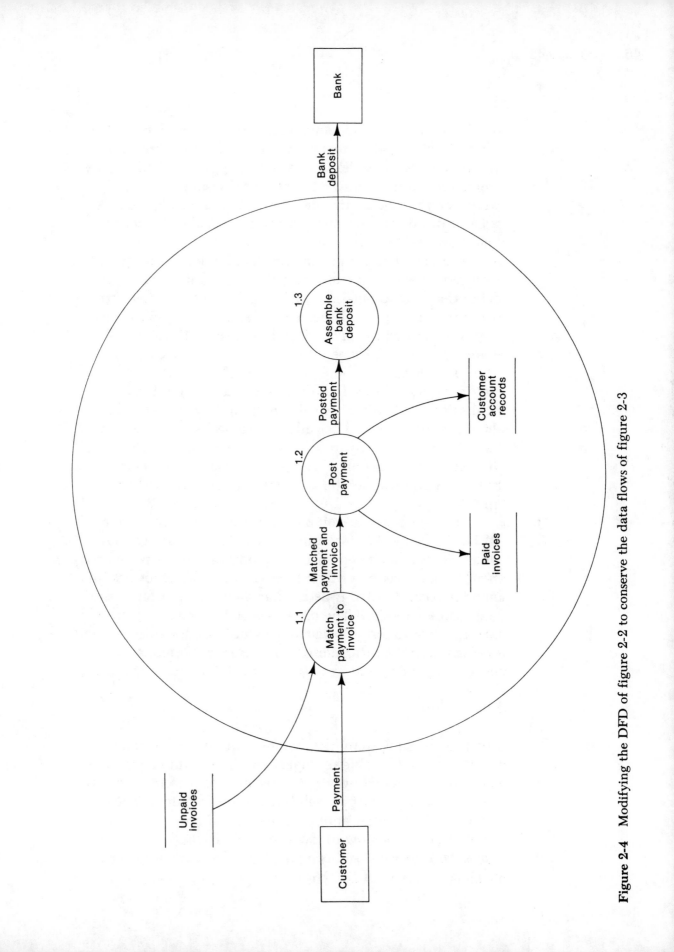

Figure 2-4 Modifying the DFD of figure 2-2 to conserve the data flows of figure 2-3

overlook areas that are complicated or that you don't understand. That would be foolish. On the contrary, do everything you need to do to understand what's happening. Write notes, create pseudocode, draw diagrams or decision tables, do whatever you feel contributes to your understanding of the system. But don't feel compelled to generate documentation for its own sake.

As a guideline for putting off detail when drawing DFDs, I suggest that you omit special-handling requirements. That's detail that can be put off until you develop the system structure charts for a system. For instance, in figure 2-2, it's likely that, occasionally, a payment will be received that can't be matched with an invoice in the UNPAID INVOICES data store. If that happens, the payment might be returned to the customer or processed in a special way and deposited in the bank. Because that's a special condition, however, I didn't show it on the DFD. After all, most processes require special-handling routines. If you draw all of these on a DFD, a diagram that would have been a clear, clean representation of a system's typical operations becomes a tangled, confused mess.

I also suggest that you omit control factors from your DFDs. In figure 2-2, for example, I didn't attempt to show what causes a bank deposit to be produced from a posted payment (or a collection of them) or when a bank deposit should be prepared. That too is detail that should be put off until a later phase of design. The proper place for presenting controls is in the description of what the lowest-level processes do, and I defer that until I start designing the new system.

Discussion

For a simple subsystem like the one in figures 2-1 through 2-4, you should be able to understand how it functions with or without DFDs. However, the techniques presented in this chapter will work just as well for a complex system as for a simple one. And most applications aren't simple. In the next chapter, you'll see how to develop a set of DFDs for a system that is both realistic and complicated. Then, you'll better appreciate the value of DFDs.

Terminology

data flow diagram
DFD
terminator
process
data flow
data store
expanding a process
levelling a DFD
conserving data flows
child process
connecting processes

Objectives

1. Name and describe the four parts of a data flow diagram.

2. Explain what is meant by these phrases:

    levelling a DFD
    conserving data flows
    connecting processes by using data stores

3. Give practical guidelines for naming DFD components
    and numbering the processes in a set of DFDs.

4. Describe two general guidelines you should follow when
    developing DFDs.

Chapter 3

How to describe an existing system using an analysis data flow diagram

 Before you can change an existing system, you need a complete understanding of how it works. In this chapter, I'll show you how to gain that understanding by creating an analysis DFD that describes the system. Developing an analysis DFD not only gives you a good understanding of the old system, but it also gives you a solid starting point for modelling the new system.

When you create an analysis DFD for a system, you do so in three steps. First, you create a DFD that describes the context of the development project. Second, you expand this into an analysis DFD that describes the system. Third, you try to simplify the analysis DFD so it is as easy to understand as possible. The DFD then becomes input to the next phase of system design, modelling the new system.

The system examples that I'm going to use in this chapter are based on a system I developed for a commodity brokerage firm. This system will be used as a basis for illustration throughout the remaining chapters in the book, so the time you spend learning about it now will not be wasted. Although the actual system is more complicated than the examples illustrate, the subset I present illustrates the central functions of the system. As a result, you should feel that you're looking at a real system. In it, you can see the complex interrelationships between the parts of a system.

Although I could have chosen a simpler system to illustrate our development methods, I didn't think a simpler system would serve my instructional purpose. Somehow an example of a system that prepares a seven-course dinner or fixes a flat tire completely misrepresents the process of system development. And examples of tiny subsystems do too. In my opinion, you have to see a method applied to genuine systems problems if you really want to understand it.

On the other hand, you shouldn't feel that you have to understand everything I tell you about the brokerage system. Its purpose is to help present our development methods. So as long as you see how our methods are used to analyze, design, and implement the brokerage system, you don't have to become an expert on commodity brokerage.

As you develop an analysis DFD, you should keep in contact with the users (you almost have to) so there will be no surprises later on. And you should put off detail as much as possible to later phases of development. This, of course, is in keeping with our general approach to system development. With that as background, let me now show you how to create an analysis DFD for an existing system.

Step 1: Create the context DFD

Figure 3-1 is a *context data flow diagram*, or *context DFD*, for the brokerage system. This is a special kind of DFD that defines the system under study. In other words, the context DFD defines the context of the development project. As a result, each of the four data flows in figure 3-1 is an input to or an output from the brokerage system itself.

Getting started   When I began the systems project for the brokerage firm, I knew little about their business. All I knew was that the firm buys grain and feed commodities from farmers and sells them, ideally at a profit, to buyers (mills, feed lots, or other farmers). Obviously, I had a lot to learn.

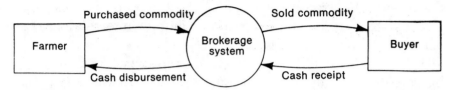

Purchased commodity   Sold commodity

Farmer   Brokerage system   Buyer

Cash disbursement   Cash receipt

Figure 3-1   The first draft of the context DFD for the brokerage system

But I began by drawing the context DFD in figure 3-1. It shows everything I knew about the system when I started the project.

At my first meeting with the staff of the brokerage firm, I began my investigation using the DFD in figure 3-1 as a basis for discussion. I thought I couldn't be wrong with such a simple diagram, but I was. First of all, I found out that the users didn't call the farmers from whom they bought commodities farmers (even though that's what they are). They call them "sellers." Well, farmers made more sense to me, but one of the points you need to keep in mind when you work on a system development project is that you should speak the language of your users. So if they wanted to call their farmers sellers, that was fine with me. I scratched FARMER off my context DFD and wrote in SELLER.

My next adjustment was more substantial. Just by looking at the context DFD, the users could tell that I didn't really understand how their business operated. Typically, the brokerage firm didn't take physical possession of a commodity in the transfer from seller to buyer. Instead, it arranged with a carrier (a trucking firm or a railroad) to move the commodity directly from the seller to the buyer. All the brokerage received was the documentation associated with the delivery.

Well, that's two changes to my first context DFD in as many minutes. So I redrew the DFD as shown in the top part of figure 3-2. As you can see, I scratched out the data flows called PURCHASED COMMODITY and SOLD COMMODITY. Then, I drew in a third terminator called CARRIER and added three new data flows: one from CARRIER to BROKERAGE SYSTEM called DELIVERY DOCUMENTS, one from SELLER to CARRIER called PURCHASED COMMODITY, and one from CARRIER to BUYER called SOLD COMMODITY.

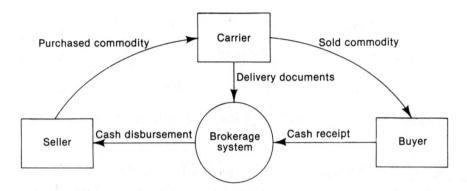

Corrected, but "improper," context DFD

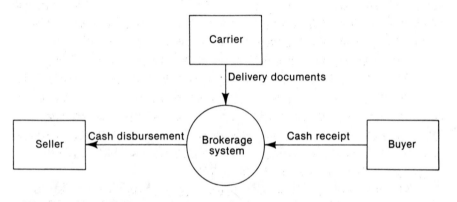

Corrected context DFD

Figure 3-2   Correcting the first draft of the context DFD
for the brokerage system

Unfortunately, the top part of figure 3-2 is "improper" in terms of context data flow diagrams. I say improper because there is no reason to include data flows that are outside the diagram's sphere of interest. In this case, all that's of interest are the data flows that pass into or out of BROKERAGE SYSTEM. The data flows I added between SELLER and CARRIER and between CARRIER and BUYER are unnecessary. I drew them in while talking with the users just to show them I understood what they were telling me as they corrected my first DFD. After we agreed that the diagram represented their business, I removed these data flows. The corrected context DFD is in the bottom part of figure 3-2.

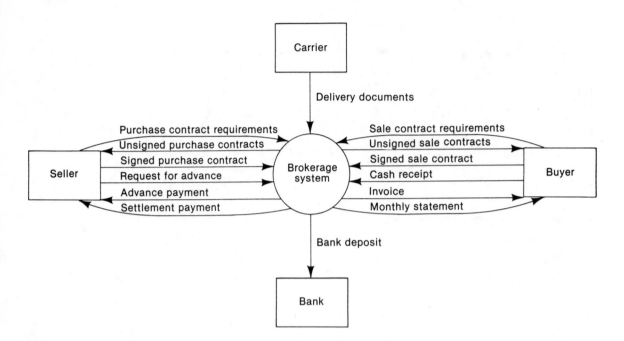

Figure 3-3   The final context DFD for the brokerage system

Identifying inputs and outputs   One of the first things you should do during the early stages of analysis is to account for all of the existing system's inputs and outputs. You do this by interviewing users and collecting samples of all source documents and output documents that you can get your hands on. In a short time, then, I discovered that I hadn't provided for some critical items in the context DFD. Among other things, the users pointed out that I had omitted:

> contracts
> invoices
> bank deposits
> monthly statements
> payments to sellers

When I added these inputs and outputs to the context DFD, it looked like the one in figure 3-3. But let me explain the additions.

Contracts, both with sellers and buyers, are made based on the contract requirements negotiated with the sellers and buyers. In figure 3-3, these inputs are named PURCHASE CONTRACT REQUIREMENTS and SALE CONTRACT REQUIREMENTS. When the requirements are in, a three-part contract is typed, and two of those parts are sent to the seller or buyer (an output). These outputs are named UNSIGNED PURCHASE CONTRACTS and UNSIGNED SALE CONTRACTS. The seller or buyer then signs and returns one of those copies (an input). These inputs are named SIGNED PURCHASE CONTRACT and SIGNED SALE CONTRACT.

It certainly makes sense to include the documents related to accounts receivable in the system (invoices, monthly statements, and bank deposits), but I just didn't think of them when I drew my first DFD. As a result, I had to add three output data flows to my context DFD: INVOICE, MONTHLY STATEMENT, and BANK DEPOSIT. I also had to add a terminator called BANK to the DFD.

Finally, the users explained to me that payments are made to sellers for two reasons. First, they issue payments for commodities purchased (called a settlement payment and included in my original context DFD as the data flow CASH DISBURSEMENT). Second, they issue an advance payment in expectation of later deliveries. As a result, I replaced my original output data flow CASH DISBURSEMENT with two others, SETTLEMENT PAYMENT and ADVANCE PAYMENT. Since an advance is issued only when a seller asks for one, I added another input data flow to my context DFD to represent such a request. It's called REQUEST FOR ADVANCE.

At this point, I had identified all of the inputs and outputs of the brokerage system. As you can see, the resulting context diagram is fairly complex, but it's not overwhelming. And I think this is typical of most of the systems you are likely to deal with. In most cases, there just aren't that many inputs and outputs that cross the system's boundary.

On the other hand, a context diagram like the one in figure 3-3 isn't particularly useful. As you'll see in a moment, though, it's a starting point for developing the analysis DFD for the system.

Step 2: Create the analysis DFD

In chapter 1, I introduced you to the *analysis DFD*. It's the major output of the first phase of system development. It will consist of a top-level DFD that represents the critical processes in the system under study. As you will see, this top-level DFD may require more than one page for its diagrams. When it's finished, the analysis DFD is used as a basis for modelling the new system. As a result, it's critical to the success of a development effort. With that as background, let me show you how to create an analysis DFD using a context DFD as a starting point.

Getting started   To start an analysis DFD, you work from the context DFD, but you look at it from the inside, not from the outside as I have up to now. At this time, then, your attention is on what the inputs are transformed into, how the inputs are processed, what the outputs are created from, and how the outputs are prepared. In short, you must expand, or level, the context DFD until you have an accurate description of the existing system. When you're done, part of this description will be the top-level, or analysis, DFD.

To understand how you start this expansion, consider figure 3-4. This is an incomplete expansion of the BROKERAGE SYSTEM process in figure 3-3. To create it, I drew a large circle representing the brokerage system. Within it, I drew small circles at the end of each data flow or group of obviously related data flows (such as ADVANCE PAYMENT and REQUEST FOR ADVANCE). As best I could, I also numbered and named each of the processes represented by the circles. For instance, I gave the name ISSUE PURCHASE ADVANCES to process 9 because it will issue advance payments based on requests for advances.

My next task is to complete the DFD started in figure 3-4. I'll do this by expanding each of the nine processes I've started. When I've expanded all of the processes, I should be able to see the relationships between the processes in figure 3-4. Then, I can complete this DFD so all processes are connected by data flows. If they aren't, I'll continue to work with the DFD until it's logically correct.

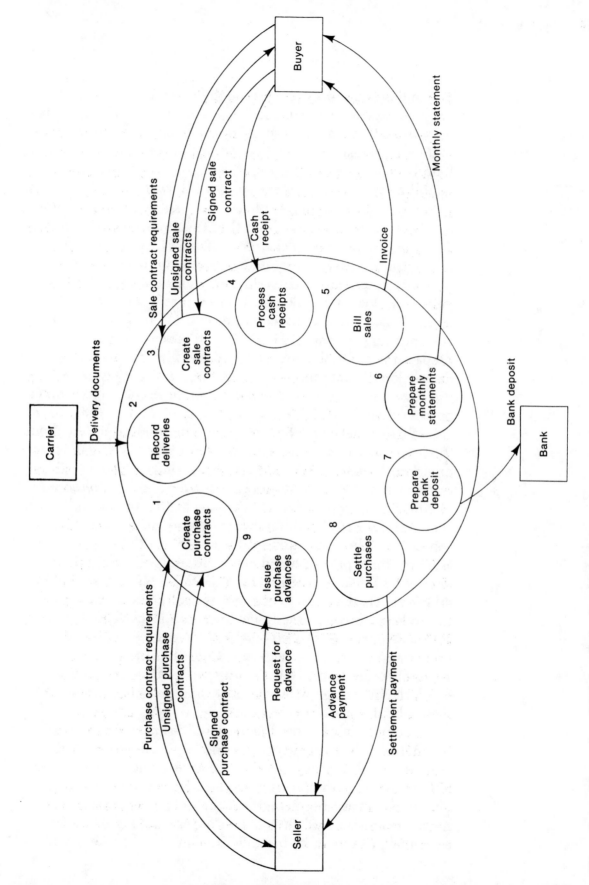

Figure 3-4 The analysis DFD for the brokerage system (version 1)

Expanding the processes   Although you can start with any of the processes you've identified, I recommend that you begin with those that start a series of transactions. If a business follows a typical sequence of events, it makes sense for you to follow it too. Since all of the brokerage transactions started by establishing a contract, I started with the contract processes.

As you prepare to complete a process, you will often have to get more information about that function of the business. So again you meet with the users. As you identify the inputs and outputs of the process under study, you begin to examine how they're processed. In addition, you should focus on the path each document takes as it passes through the system. For inputs, trace them to their final disposition to determine how current filing systems are organized. If documents are prepared in multiple parts, be sure you understand the disposition of each part. As you gather information for one process, then, you will be picking up information about other processes.

When I met with the users to get more information about the preparation of contracts, processes 1 and 3, I began to understand these processes. In brief, a contract is made as a result of a conversation between a broker and a representative of either a buyer or a seller. Simply put, the broker will buy a commodity, then try to sell it for more, or will sell a commodity, then try to buy it for less. Since the process is identical for a purchase and a sale, I'll only illustrate the expansion of CREATE PURCHASE CONTRACTS. Just keep in mind that creating a sale contract works the same way.

When a broker closes a deal with a seller, a purchase contract is prepared based on the terms agreed to by the broker and the seller. The terms state what commodity is being purchased, in what quantity, at what quality, how much the brokerage will pay for it, and how the commodity should be shipped.

The purchase contract is prepared in triplicate. The original and the second copy are mailed to the seller, who keeps one and signs and returns the other. The copy returned to the brokerage by the seller is placed in a numeric-sequence purchase contract file. The third copy of the purchase contract, printed on card rather than paper stock, has a ledger-card grid printed on its back. This enables the brokerage to

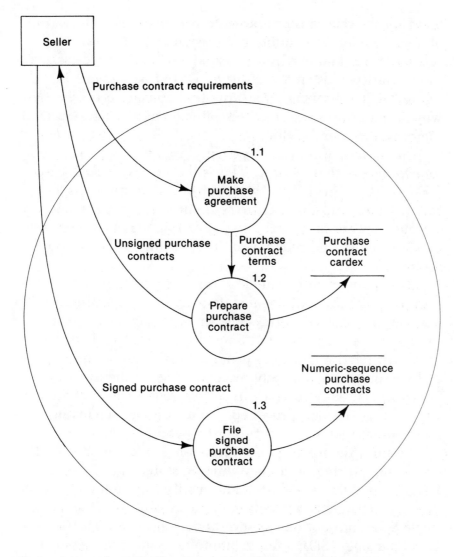

Figure 3-5   The DFD for CREATE PURCHASE CONTRACTS

keep track of deliveries and payments on that contract. These purchase-contract copies are kept in cardex file drawers organized by commodity.

With this information in mind, I was able to expand the process CREATE PURCHASE CONTRACTS in figure 3-4. The resulting DFD is shown in figure 3-5. Based on your introduction to DFDs in chapter 2, the meaning of this expanded DFD should be apparent to you.

Showing the relationships between processes A critical part of analysis is understanding how a system's processes are related to one another. As you expand a DFD's processes, you show relationships on the top-level DFD (like the one in figure 3-4) by drawing data flows that connect processes by way of data stores. Let me illustrate by continuing to expand the brokerage firm's DFD.

In general, if a process has only inputs or only outputs, something is wrong with it. This is the case with RECORD DELIVERIES, which has an input but no outputs. As a result, I met with the users to find out what actually happens as deliveries are recorded. As you read my description of this process, you should relate it to the resulting DFD as shown in figure 3-6.

First, documents related to a specific delivery are matched, although all documents related to a specific delivery may not be received at the same time. If they are, they're simply clipped together and immediately processed. If they aren't, the clerk looks in a "pending" file that contains all the delivery documents already received to see if the necessary items have already come in. If so, the related documents are clipped together and processed. If not, the documents just received are themselves put into the pending file.

Second, the matched delivery documents are processed by preparing, posting, and distributing a statement of deliveries (SOD). An SOD lists weight and quality information from the delivery documents for a given quantity of a commodity. A single SOD may list one or several deliveries, but all that appear on a single SOD must relate to the same purchase contract (and logically, be from the same seller). They must also relate to the same sale contract (and logically, be to the same buyer). Also, they must have been shipped by the same carrier.

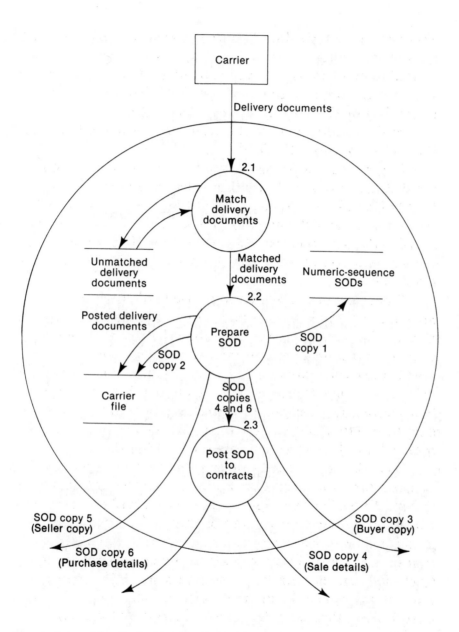

Figure 3-6   The DFD for RECORD DELIVERIES

When I analyzed the distribution of the copies of the SOD, I found that one is prepared in six parts. These parts are distributed like this:

1 (original)	numeric SOD file
2	filed with delivery documents in carrier file
3	with invoice to buyer
4	into sale contract cardex file
5	with check to seller
6	into purchase contract cardex

So these are the outputs of this process. In addition, this distribution list indicates that the RECORD DELIVERIES process requires two new data stores: (1) a numeric-sequenced SOD file and (2) a carrier file.

Once I had this information, I created an expanded DFD for the RECORD DELIVERIES process like the one in figure 3-6. As you can see, its input balances with what is shown in figure 3-4. But its newly identified outputs (the third, fourth, fifth, and sixth copies of the SOD) do not. This imbalance forces me to make a critical move in the development of the top-level DFD of figure 3-4. I have to connect processes using data flows.

The SOD copies that will accompany the check to the seller (copy 5) and the invoice to the buyer (copy 3) will have to go to the processes that perform those functions. That's simple enough. The disposition of copies 4 and 6, however, relates to data stores (the cardex files) that are internal to the processes CREATE PURCHASE CONTRACTS (figure 3-5) and CREATE SALE CONTRACTS. As a result, I have to remove those data stores from their original processes to make them available to other processes. This does not indicate that my earlier view of the system from the perspective of contract creation was wrong, but only that it was incomplete. At any rate, figure 3-7 shows my changes to the top-level DFD to accommodate these relationships.

If you compare figures 3-4 and 3-7, you'll notice a couple of differences in addition to the connections I just described.

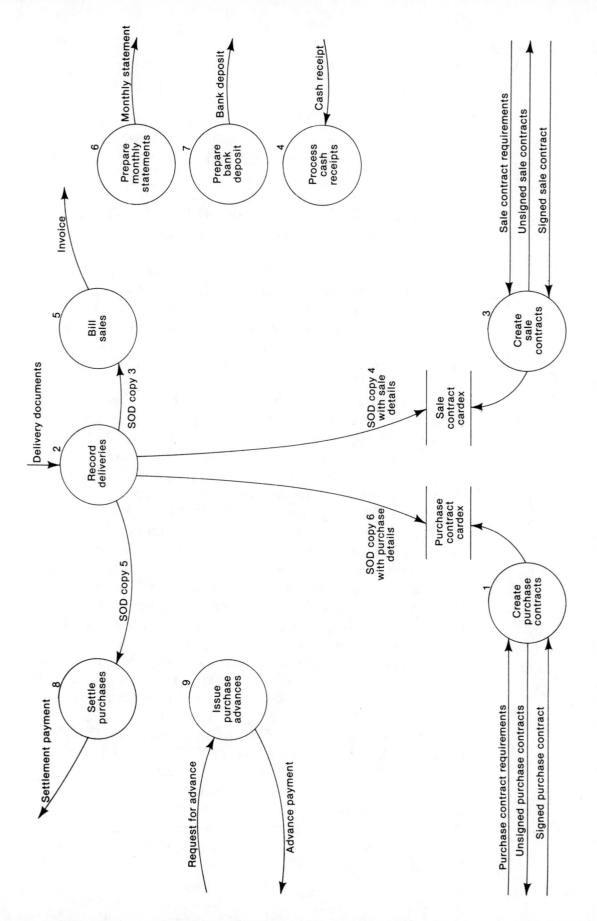

Figure 3-7 The analysis DFD for the brokerage system (version 2)

In truth, I drew figure 3-4 just to illustrate my thought processes as I started to create my detailed DFD. Now that I've
done that, I can change the DFD so it's easier and more practical to work with.

First, figure 3-4 really doesn't need to include the large
circle that stands for the entire BROKERAGE SYSTEM. After
all, that's what the study is now focusing upon. So I omitted
this circle in figure 3-7.

Second, you can omit terminators. Just be sure you know
where inputs are coming from and where outputs are going
to. In fact, in most systems, you'll find that so many data
flows enter and leave the overall system that to connect them
to the proper terminators results in a confusing DFD. Accordingly, I also omitted terminators from figure 3-7.

Completing the first draft of the analysis DFD   To complete
my analysis DFD, I continued to examine the processes I
drew at the ends of the context DFD's data flows. As I expanded the remaining processes, I took care to insure that the
data flows were conserved. Also, I paid particular attention to
the sharing of data stores by two or more processes.

I expanded the process SETTLE PURCHASES next. As
you read the description that follows, you can use the
resulting DFD in figure 3-8 as a guide. When I investigated
this process, the users explained to me that the fifth copies of
SODs are held until someone decides to pay a seller for commodities purchased from him. When a payment is to be
made, all the SODs to be paid that are related to a specific
contract are gathered. Then, referring to the pricing terms in
the purchase contract in the cardex file, the settlement
amount to be paid is calculated. Finally, the payables clerk
writes the check and records the payment in the cardex file.

When I expanded SETTLE PURCHASES, I realized that
I had to add a pair of data flows to my top-level DFD for the
system. Both of these flows connect SETTLE PURCHASES
and the PURCHASE CONTRACT CARDEX data store. The
first one indicates the retrieval of pricing information from
the data store. The second one, flowing the opposite direction, indicates the recording of a payment. You can look
ahead to figure 3-11 if you want to see how these data flows
look on the top-level DFD.

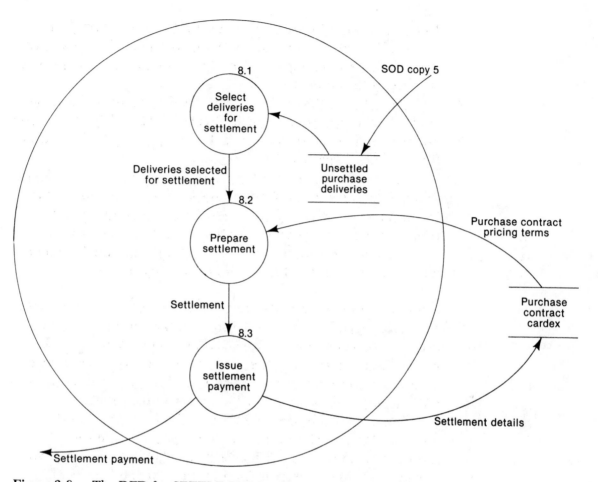

Figure 3-8   The DFD for SETTLE PURCHASES

At this time, I turned my attention to the BILL SALES process. The resulting DFD is shown in figure 3-9. At first, when I talked to the users, it seemed just like SETTLE PURCHASES. An SOD copy is placed in an unbilled delivery file. From that file, SODs are selected for invoicing. The selection process here is simpler than on the purchase side, however. Since billing helps get money into the bank as rapidly as possible, all SODs are billed every day unless there are exceptional conditions.

As I continued my investigation into the process, I found that the SODs that are to be billed are grouped according to contract. Next, the sale contract cardex file is consulted to determine the pricing terms that should be used. Then, a

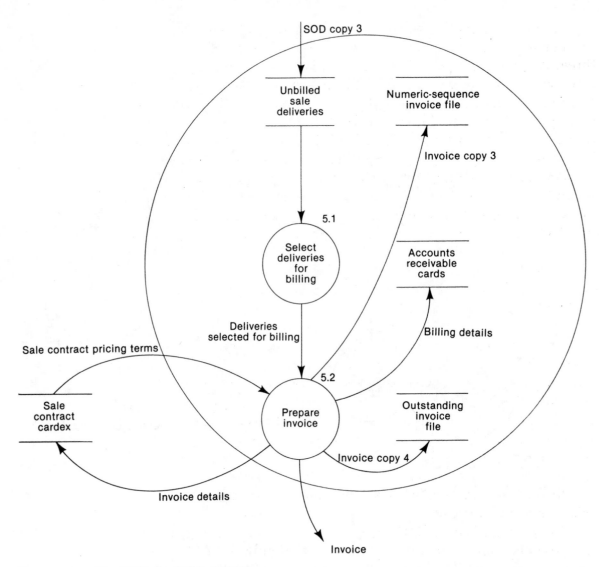

Figure 3-9   The DFD for BILL SALES

four-part invoice is typed. After the amount being billed is recorded on the sale contract cardex and on an accounts receivable card, the four parts of the invoice are distributed. Two parts go to the buyer along with the copies of the SODs that were used to prepare them. The third copy of the invoice is placed in a numeric-sequence invoice file. The fourth copy is placed in an outstanding invoice file that's used for accounts receivable.

If you compare the DFDs in figure 3-8 and 3-9, you may wonder why one contains two processes and one contains three. After all, these processes are comparable in terms of complexity. My answer is that I can expand almost any process in more than one way. Yes, I could have drawn three or more processes within the expanded DFD for BILL SALES. And I could have expanded process 5.2 in figure 3-9 one more level down. But what I'm trying to do is to draw a DFD that *I* understand. As soon as I understand it, I stop trying to expand the process further. In this case, the BILL SALES process was easier for me to understand than the SETTLE PURCHASES process, so I used fewer process circles in my expansion of BILL SALES. When you level the processes of a top-level DFD, then, remember that the resulting DFDs are your working papers. If you understand them, they're adequate.

In any event, the last process on my top-level DFD that was complicated enough to expand in detail was PROCESS CASH RECEIPTS. Its DFD is shown in figure 3-10. In brief, when a payment is received in the mail, it's matched to the corresponding invoice or invoices in the outstanding invoice file. After the payment is matched with its invoices, the invoices are marked paid and filed permanently with the signed copies of their related sale contracts in the numeric-sequence sale contracts file. Information about the payments is also recorded at this time on the accounts receivable cards and the sale contract cardex.

When I finished the DFD in figure 3-10, I realized that I had to make some changes to my top-level DFD of figure 3-7. To make all of the data stores required by PROCESS CASH RECEIPTS available to it, I had to remove them from the processes that contained them. Specifically, I removed the data stores called ACCOUNTS RECEIVABLE CARDS and OUTSTANDING INVOICE FILE from the process called BILL SALES. I also removed the data store called NUMERIC-SEQUENCE SALE CONTRACTS from the process called CREATE SALE CONTRACTS.

Figure 3-11 shows the top-level DFD with the changes required by the expansion of SETTLE PURCHASES, BILL SALES, and PROCESS CASH RECEIPTS. In addition, I added a couple of other data flows that I discovered during

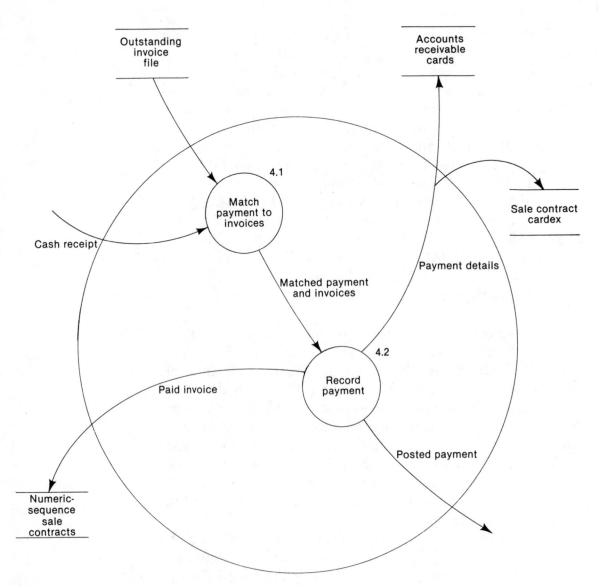

Figure 3-10 The DFD for PROCESS CASH RECEIPTS

my conversations with the users. First, the process ISSUE
PURCHASE ADVANCES causes an entry to be made in the
PURCHASE CONTRACT CARDEX. Second, the input of
PREPARE MONTHLY STATEMENTS is the data store
ACCOUNTS RECEIVABLE CARDS. Finally, the data flow
POSTED PAYMENT goes from PROCESS CASH RECEIPTS

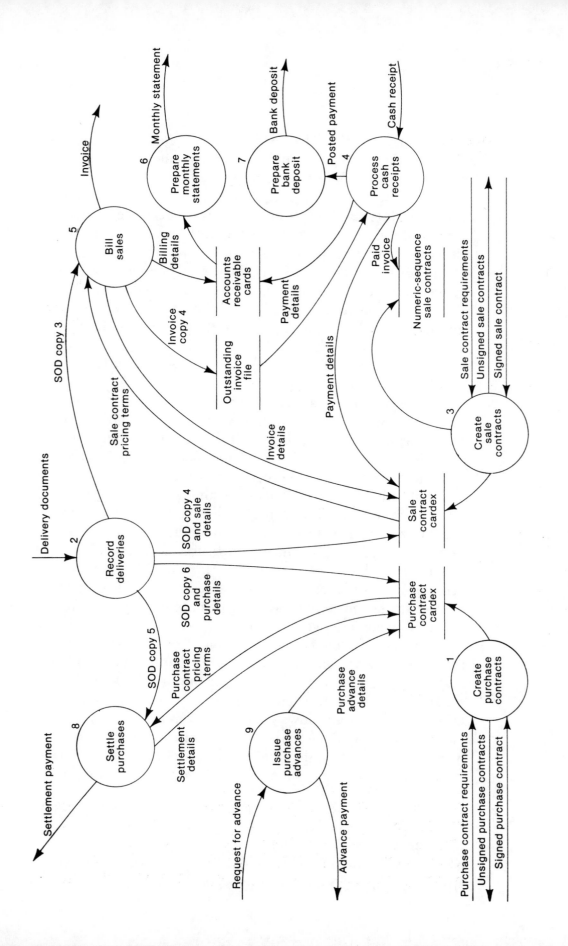

Figure 3-11 The analysis DFD for the brokerage system (version 3)

to PREPARE BANK DEPOSIT. At this point, I've accounted for all of the inputs and outputs on the context DFD of figure 3-3. But I'm still not done with my analysis DFD.

Showing "internal" inputs and outputs   This topic heading may well sound like a contradiction in terms. However, I think it aptly describes a class of data flows that you need to recognize as you develop DFDs for actual systems. Using the conceptual model I've described, the analysis DFD represents the overall system under study, and the data flows that enter and leave it either come from or go to terminators. In most systems, however, you'll find inputs and outputs that don't come from or go to terminators. I call these *internal inputs* and *internal outputs*.

In the brokerage system, for example, after I investigated all of the processes based on the external inputs and outputs I had identified, I asked the users if they used any internal documents or reports. After some discussion, we concluded that there were three. The first, a "mark-to-market" report, is a complicated document that shows the amounts they will bill and pay for commodities contracted to be bought and sold compared to their values on the commodities market. The second, an aged-trial-balance report, lists all outstanding accounts receivable items by customer and classifies them according to how long they've remained unpaid. The third, a daily management report, summarizes accounting data from the firm's general ledger system.

Somehow, of course, the creation of these three reports needs to be shown on the analysis DFD. To do this, I suggest that you consider the management of the firm under study as a terminator in itself. Then, processes can be added to the DFD that show data flows going to and from the management terminator. Even if you omit terminators from your working DFDs, there shouldn't be any confusion about where internal data flows go as long as you give them meaningful names.

To show the preparation of the first two of these internal reports, I added two processes to the top-level DFD as shown in figure 3-12. The first, PREPARE AGED-TRIAL-BALANCE REPORT, draws information from the OUTSTANDING INVOICE FILE data store. The second,

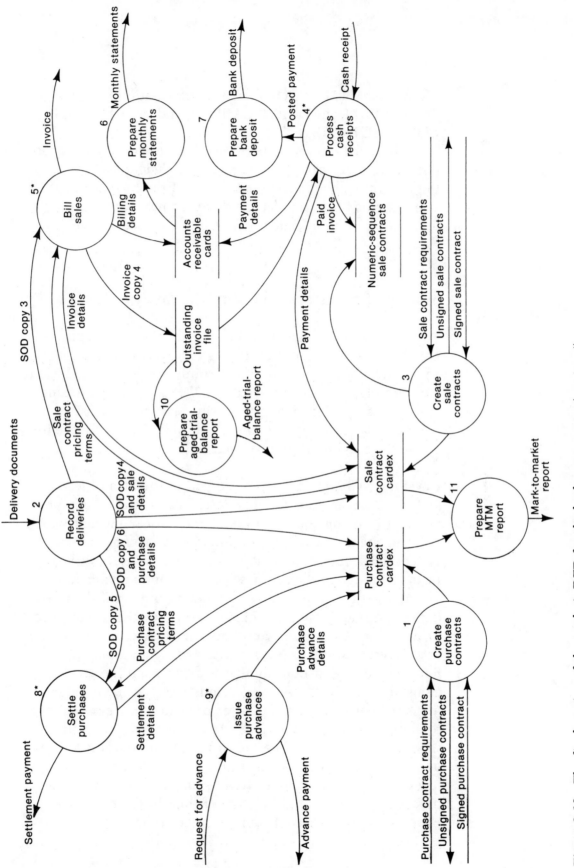

Figure 3-12 The final version of the analysis DFD for the brokerage system (version 4)

PREPARE MTM REPORT, draws information from the PURCHASE CONTRACT CARDEX and the SALE CONTRACT CARDEX.

The third internal report, the daily management report, is created solely from the firm's general ledger system. Since the general ledger system isn't a part of the brokerage system I've described thus far, let me defer for just a moment a description of how I integrated it into the analysis DFD.

Step 3: If necessary, simplify the analysis DFD

You can see now that for all but the simplest of systems, a DFD will be complex. In fact, it's relatively easy to create a complicated diagram, even for a simple system. And if you're not careful, you can complicate a DFD further by crossing lines and arranging elements in ways that aren't sensible. At this stage of your analysis, then, you should consider the use of the simplifying techniques that follow.

Levelling   Chapter 2 introduced you to the idea of *levelling*, so I won't cover it again. However, I can't stress the importance of levelling enough since it's the most important way you can simplify a DFD. Levelling allows you to break a system down into manageable parts while it prevents you from losing track of how those parts relate to one another.

If you draw a process with a large number of data flows coming into and going out of it, the whole DFD becomes less meaningful. As a result, you should try to keep the number of data flows entering or leaving a process to a reasonable number. Just what that number is depends on the specific circumstances of each case, so you must use your own judgement. In figure 3-12, the most complicated processes have a total of six inputs and outputs. As a rule of thumb, you should try to keep the number of inputs and outputs for any one process to a total of six or less. The critical factor, however, is understanding. If you can easily understand a process, you can exceed this rule of thumb. On the other hand, if a process is confusing, it ought to be levelled.

In any event, if your analysis DFD contains processes with an excessive number of data flows entering or leaving them, you should level those processes. As an obvious example, the brokerage system described in the context DFD in figure 3-3 needs to be levelled if it's to be meaningful. If a DFD is complicated enough, you may need to take it down two or more levels. When you have levelled a DFD to a point where it's easy to understand and use, you have a *levelled DFD*. In most cases, of course, the levelled DFD will consist of several sheets of paper: one for the top-level DFD and one for each of the process expansions.

Layering   Another way you can simplify a DFD is to use *layering*. A *layered DFD* is a set of drawings that's based on the fact that most systems are too complicated to be represented by a two-dimensional drawing. In a layered DFD, you draw two or more DFDs, but they have some processes in common. For complex systems, this is a practical way of simplifying an analysis DFD. In contrast, if you try to draw a single DFD with all of the interconnections that exist in real systems, you'll often find that the result is unmanageable, if it's possible to draw at all.

In the brokerage system, for example, one interface I've neglected thus far is the connection between the processes in figure 3-12 and accounting. If I tried to add this connection to the DFD in figure 3-12, it would probably end up a mess. But I can add it quite easily by using layering.

Specifically, the processes in figure 3-12 that interface through a posting process to the accounting system are:

> SETTLE PURCHASES
>
> ISSUE PURCHASE ADVANCES
>
> BILL SALES
>
> PROCESS CASH RECEIPTS

I challenge you to try to draw a readable DFD on one page that shows these relationships. Adding a new process like PERFORM ACCOUNTING TASKS to the DFD in figure 3-12 would severely complicate the diagram.

The layered alternative is to draw a related diagram that duplicates the appropriate processes from figure 3-12 and shows only the data flows from them into the new process. To do this, imagine a new process called PERFORM ACCOUNTING TASKS that's in a "layer" above the plane of the paper figure 3-12 is printed on. Figure 3-13 gives a perspective view of what you should be imagining. Keep in mind, however, that I include this drawing just to let you get a better mental image of what I'm talking about.

Figure 3-14 illustrates the type of DFD you'll create when you use layering. It's the top layer for the brokerage firm's system. When you use layering, you view the layers as parts of a single analysis DFD. As a result, you don't show all of the data flows for a process on both layers. Note, for example, that I didn't add the data flows that go to the process PERFORM ACCOUNTING TASKS to my original DFD in figure 3-12 because they would complicate the diagram. After all, the purpose of layering is to simplify the analysis DFD.

When you create a layered DFD like this, you must be sure you use the same names for the processes you duplicate from one layer to the next. And you can see that I've done so in the layers of figure 3-12 and 3-14. It also helps to show which processes are duplicated from one layer to another. In figures 3-12 and 3-14 I've identified the processes that appear in both layers with an asterisk next to the process number (processes 4, 5, 8, and 9). You can adopt this convention if you like. Or you can adopt your own convention like identifying the common processes by using another symbol or a color.

As I've said, you don't show all of the data flows for a process on each layer of an analysis DFD. So if you want to identify all of the data flows for a process, you must analyze the data flows on all of the layers of the analysis DFD. Then, if you want to have clear documentation of all of the data flows for a process, you can show them on the lower-level DFDs. That way the data flows are conserved from the layered DFD to the lower-level DFDs.

For instance, I could add output data flows to processes 4, 5, and 8 in figures 3-10, 3-9, and 3-8 to represent the data flows that go to the process PERFORM ACCOUNTING TASKS. As I see it, though, adding these data flows is optional since the main purpose of the lower-level DFDs is to

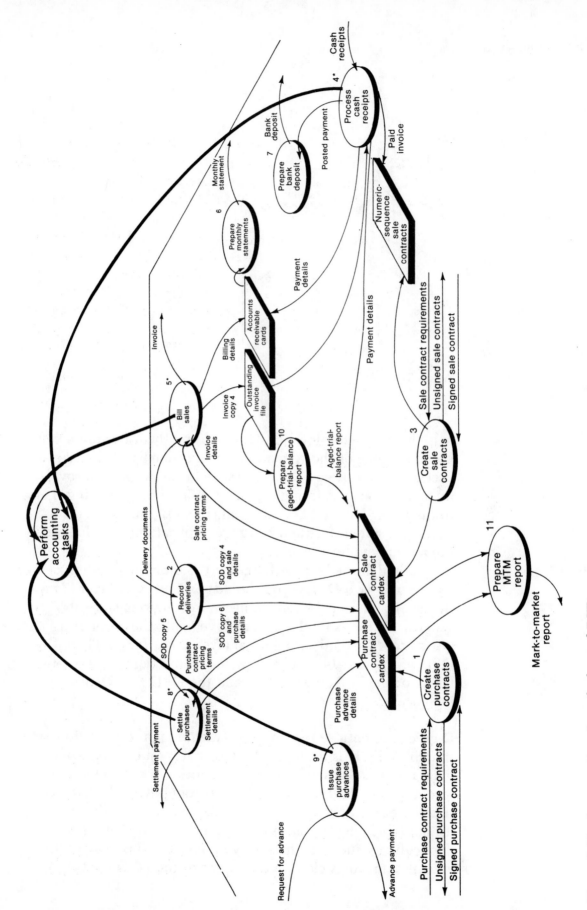

Figure 3-13 A perspective view of a layered DFD

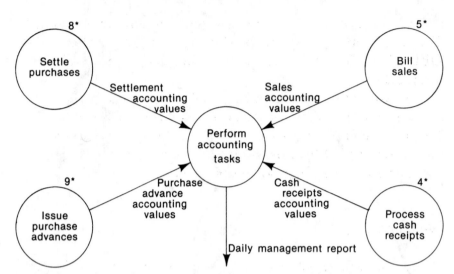

Figure 3-14   A DFD representing the top layer of the brokerage system

help create the top-level DFD. If adding them improves your understanding of the system, do so. But don't feel that it's necessary. In this case, since these data flows simply provide linkage to the accounting system, I didn't bother adding them to the lower-level DFDs. After all, my assignment was to develop a system for the brokerage applications, not for the accounting applications.

Although the layered DFD for the brokerage system consists of only two DFDs, you should realize that a layer like the one in figure 3-14 can connect several systems or subsystems. If, for example, I were developing the accounting system too, I would create an analysis DFD for it. Then, the DFD of figure 3-14 would show the connections between the DFD for the brokerage system and the DFD for the accounting system.

Duplicating data stores   Another method you can use to simplify a DFD is to duplicate data stores on one page of a diagram. This is useful when data stores are connected to several processes. In figure 3-12, for example, each of the contract cardex files is connected to five processes. In this case, it's not too difficult to show this in a single DFD without duplicating some elements. All you have to do is position the critical data stores close to the center of the DFD. But sometimes several data stores will have similar connections. Then, you can't put all of the data stores in the center of the DFD. So duplicating some of them becomes a practical alternative.

As you will see later in this book, some data stores that are used for reference can be accessed throughout a system. To show them in one place with all the proper connecting data flows would be a drafting nightmare. But duplicating the data stores can help solve the problem.

Discussion

Now, you've seen how to create an analysis DFD. At this point, the analysis DFD for the brokerage system consists of the diagrams in figures 3-12 and 3-14. These diagrams in turn become input to the next phase of system development, modelling the new system. I will also keep the expanded DFDs for each of the processes in figure 3-12, like the ones in figures 3-9 and 3-10, but they'll be less useful. Their main purpose was to help me develop the analysis DFD. Along with these DFDs, I'll keep the source documents, report samples, notes, and so on, that I've collected as part of this analysis phase. In future development phases, I'll use these as working papers to make sure I don't miss anything. But the only required documentation at the end of this phase of development is the analysis DFD.

Although I would have liked to present a more precise procedure for developing an analysis DFD, I think you can see by now that it's impossible to prescribe one. In general terms, I can say the procedure is to (1) create the context DFD, (2) create a complete version of the analysis DFD, and (3) simplify this version. But this leaves much unsaid. In order to create the complete version of the analysis DFD, you need to do a considerable amount of creative and analytical thinking on your own. It's you who must decide in what sequence to expand the processes and how to connect the processes using data flows. And all the while you must work with the users to get a complete understanding of each process.

When you're ready to simplify the complete DFD, you must realize that a tradeoff has to be made between readability and redundancy. In general, you should duplicate items if this simplifies your drawings. But you should avoid duplication that has no purpose. In other words, duplicate a data store or create a layered DFD only if it clarifies your drawings. Here again, though, I'm unable to prescribe a precise procedure for simplification. So you must make these decisions on your own.

If you use the methods presented in this chapter, I think you'll find that they're more efficient and effective than any analytical methods you've used in the past. So I think you'll realize a couple of benefits right away.

First, you'll reduce the time it takes you to analyze an existing system. For the brokerage system I've just described, it took me less than two weeks to develop the analysis DFD for the system, even though it was more complicated than the one presented in figures 3-12 and 3-14. At this point, I was ready to model the new system in preparation for designing it. In contrast, using traditional techniques, it's not unusual to spend two or three months analyzing a system that's no more complicated than the brokerage system.

Second, the analysis DFD is more useful when you start to design the new system than traditional forms of documentation are. In our method, the existing system is described on just a few sheets of paper. And this description is efficient input to the next phase of system development. In contrast, using traditional techniques, you might end up with two or more 3-ring binders full of narrative and forms that in total don't help you model the new system. In fact, traditional documentation like this works against you because it's so overwhelming that no one wants to read it. And worse, it usually contains so much useless detail that it's easy to overlook the critical aspects of the system.

In summary, I believe the use of analysis DFDs can make an important contribution to your development efforts. If you're wondering how the two or three pages of an analysis DFD can provide enough background for modelling a new system, I can only urge you to give this method a try. When we've used analysis DFDs to describe systems that we've developed, we've had no serious development problems due to misunderstandings of the old systems. And we reduced the time spent analyzing the old systems to sensible proportions.

As a result, we were able to devote the major portion of our
energies to designing and implementing the new systems.

Terminology

context DFD
analysis DFD
internal input
internal output
levelling
levelled DFD
layering
layered DFD

Objective

Given an existing system, create an analysis DFD that
describes it.

Chapter 4

How to model a new system
using a model data flow diagram

In this chapter, I'll show you how to begin the creative work of planning a new system. When you complete this chapter, you'll know how to develop a model DFD for a new system. As you'll see, a model DFD focuses upon the central data stores of the new system and the relationships between its processes. As a result, the model DFD provides an excellent starting point for the later design phases.

Although I'd like to give you a precise procedure for modelling a new system using a model DFD, I can't. As you'll see, the modelling process depends to some extent on insight, creativity, common sense, and experience. As a result, I can only give you some guidelines for this creative process. First, I'll present a general procedure for modelling a new system. Then, I'll explain how I created the model DFD for the brokerage system.

A general procedure for modelling a new system

The general procedure for modelling a new system consists of four steps. First, you identify the changes in the functional requirements for the new system. Second, you establish the context for the new system. Third, you create a context DFD for the new system. Fourth, you create the model DFD for the new system. When you're done, the model DFD becomes input to three of the design phases of our system development method.

Step 1: Identify the changes in functional requirements for the new system   As you create an analysis DFD for an existing system, you talk to the users. It's natural, then, that you'll get some ideas from them about the new system. It's also natural that you'll get some ideas of your own.

When you complete the analysis DFD, in fact, you may have a clear idea as to what the functional requirements are for the new system. If not, you'll have to meet with the users again to get their ideas about the requirements of the new system. In general, the requirements for the new system consist of the requirements of the old system plus any new requirements. Sometimes, however, the users realize that some of the functions of the old system are no longer necessary, so they drop them from the requirements for the new system.

If you work in a large shop, you may have a formal procedure for identifying the requirements of a new system. This procedure may include tasks like setting system objectives, setting performance criteria, and ranking the proposed requirements using cost/benefit analysis. As I've said before, however, these traditional techniques of analysis are beyond the intent of this book. My purpose is to focus on a development method, not on techniques for system justification.

Figure 4-1 presents the changes in functional requirements for the new brokerage system. To organize these requirements, I've divided them into operational, reporting, and inquiry requirements. I'll talk more about these requirements when I show you how I created the model DFD for the brokerage system. But let me present them briefly right now.

The operational requirements reflect the fact that the old system was error-prone. In fact, a major benefit of the new system was improved accuracy, a benefit that had the potential to justify the entire system by itself. In the previous year, for example, the company had issued checks to the wrong seller, had paid two sellers twice for the same contract, had

Operational requirements

1. Replace the time-consuming and error-prone manual operations for recording transactions (deliveries, settlements, billings, and payments) with an efficient alternative means of recording the transactions.

2. Check the lienholders at time of settlement to prevent a disbursement to the wrong seller.

3. Check advance payments at time of settlement to prevent double payments to a seller.

4. Check advance billings at time of invoicing to prevent double billings.

5. Match the purchase and sale contracts when recording a delivery to insure that:

   - the delivery is applied to the proper contracts;
   - deliveries beyond the amounts contracted for do not happen accidentally; and
   - the proper prices are used.

6. Issue all checks, whether settlement payments or advance payments, from one source to eliminate errors.

Reporting requirements

1. Prepare an aged payables report
2. Prepare a position report
3. Show profit by commodity on the daily management report.

Inquiry requirements

1. Provide an inquiry capability to be used by both operational and management personnel for all automated files. It should provide both detail and summary data for:

   buyers
   sellers
   purchase contracts
   sale contracts
   deliveries
   settlements
   advances
   invoices
   cash received
   cash disbursed

Figure 4-1   The changes in functional requirements for the new brokerage system

allowed deliveries beyond the quantities contracted, and so on. Besides that, the company was never able to reconcile its operational data with its accounting data. In short, the company needed a better system, it needed it badly, and there was no doubt that its problems justified the installation of its first computer system.

As for the reporting and inquiry requirements, they represent a sort of "wish list" requested by the managers of

the company. They wanted two new reports. They wanted
more detail on the daily management report. And they
wanted to be able to get at the information of the system in
seconds rather than minutes. In the old system, it was difficult for them to answer questions about any operational
details because most of the critical information was recorded
on cards that rarely seemed to be in the cardex file when they
needed them.

Step 2: Establish the context for the new system   After you've
identified the changes in functional requirements for the new
system, you're ready to start to model it. When you start to
model a new system, the first step is to determine what its inputs and outputs are. In other words, you must establish the
context for the new system.

In most cases, the context of the new system is the same
as the context of the old system. But sometimes the scope of
the development project is reduced after the existing system
has been described. In this case, you work from a copy of the
analysis DFD to establish the reduced context. First, identify
the processes that are to be included in the development project. Then, draw a circle around them to establish the boundary of the new system.

For example, imagine that the scope of the brokerage
system project was reduced to improving collection procedures. Then, the boundary for the new system would probably be the one outlined in figure 4-2, although it might also
include BILL SALES. In either case, what happens in contracting, recording deliveries, or settling purchases is irrelevant to the collections subsystem. As a result, you exclude
them from the development project. You've thus established
the context of the project.

Once you've established the boundaries of the new
system, the data flows that cross it become *some* of the inputs
and outputs of the new system. In addition to these, you must
add any new data flows that come from the change requirements for the new system. Viewed from the wider
perspective of the complete DFD, two of the data flows for
the collections subsystem in figure 4-2 come from another process; two go to data stores; and the remaining four either
come from or go to terminators. But from the narrower

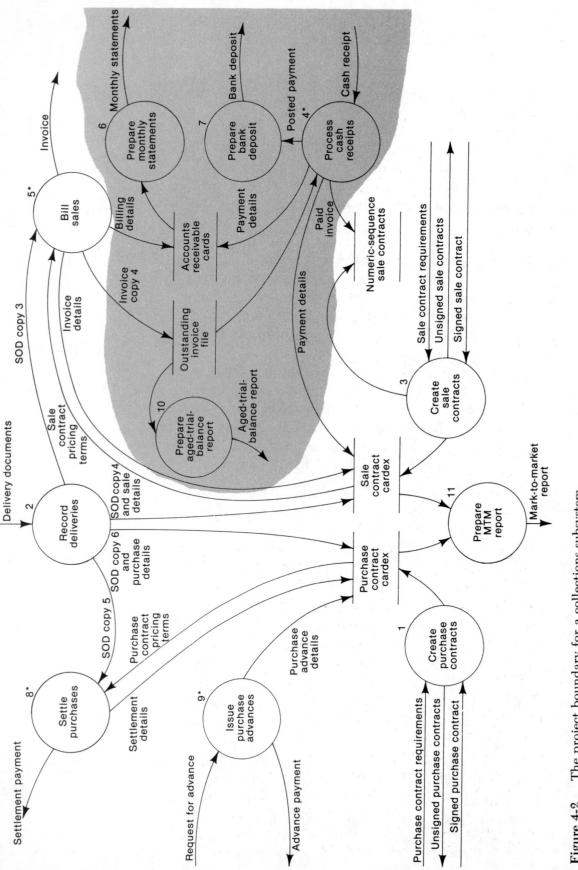

Figure 4-2 The project boundary for a collections subsystem

perspective of the collections subsystem, these facts are irrelevant. In effect, all of the data flows that pass through the boundary for the collections subsystem appear to it to come from or go to terminators. They're the inputs and outputs for the context you've established.

When I developed the brokerage system, almost all of the system described by the analysis DFD was included in the context for the new system. Excluded was the accounting subsystem because a software package was available for these functions. Unfortunately, this software package didn't provide for the preparation of the daily management report. So I was also responsible for its preparation.

With this in mind, I established the context of my new system as all of the functions represented by the analysis DFD of figure 3-12 and 3-14, plus the additional functional requirements listed in figure 4-1. Knowing this, I could create a context DFD for the new system.

Step 3: Create a context DFD for the new system   To create a context DFD for a new system, you need to know all of the inputs and outputs it requires. As a result, I listed all of the inputs and outputs for the new brokerage system as shown in figure 4-3. Because the old and the new systems have the same boundary, this list includes all of the old system's inputs and outputs (data flows) as represented by the analysis DFD of figure 3-12 and 3-14. It also includes the data flows that are required by the functional changes for the new system.

After you've listed all of the new system's inputs and outputs, you should identify any output data flows that are prepared by extracting information from a file. Similarly, you should identify any input data flows that are filed but not used again by the system. If you look at the right-hand column in figure 4-3, you can see that I identified several data flows like this. Then, when I create the context DFD, I can omit these inputs and outputs since they have only a trivial effect on the design of the new system. This will simplify the context DFD, and it will also simplify the resulting model DFD. As you develop the system, of course, you need to provide proper access paths for the extracted data flows, but I'll cover that in chapter 7. Right now, though, you must

Process	Data flow	Input/Output	Exclude?
1	Purchase contract requirements	I	
	Signed purchase contract	I	Yes—just filed away
	Unsigned purchase contracts	O	
2	Delivery documents	I	
3	Sale contract requirements	I	
	Signed sale contract	I	Yes—just filed away
	Unsigned sale contracts	O	
4	Cash receipt	I	
5	Invoice	O	
6	Monthly statement	O	Yes—extraction
7	Bank deposit	O	
8	Settlement payment	O	
9	Request for advance	I	
	Advance payment	O	
10	Aged-trial-balance report	O	Yes—extraction
11	Mark-to-market report	O	Yes—extraction
12	Daily management report	O	Yes—extraction
New	Position report	O	Yes—extraction
New	Aged payables report	O	Yes—extraction
New	Inquiry to replace cardex files	O	Yes—extraction

Figure 4-3   The input and output data flows for the new brokerage system

simplify the context DFD if you want to be sure to emphasize the critical data flows and processes as you model the new system.

Based on your list of inputs and outputs for the system, then, you prepare the context DFD just as you would if you were analyzing an existing system. For the new brokerage system, my context DFD is shown in figure 4-4. If you compare this with the context DFD for the old system in figure 3-3, you can see that the new system DFD is less complex. Since you're deliberately trying to simplify the context of the new system, this is as it should be.

Step 4: Create the model DFD for the new system   When I show you how I created the model DFD for the brokerage

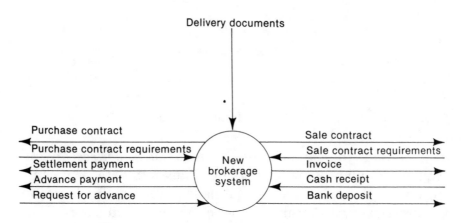

Figure 4-4   The context DFD for the new brokerage system

system, I'll give you specific ideas for creating model DFDs. Right now, though, I'd like to give you the general idea of what you should be trying to do as you create the model DFD for a new system. As I see it, there are two principles that are critical to the success of this phase of development.

The first principle is: *Keep the model DFD simple.* As you begin to model a new system, it's essential that you emphasize critical factors and avoid being distracted by details. If you don't strip away the nonessential details, you can easily make a simple project complex and a complex one impossible. As you develop a model DFD, then, your goal should be to omit processes and data stores that don't have an impact on the selection and arrangement of critical DFD components.

Figure 4-5 lists the DFD components you can safely omit when you create model DFDs. I'm not suggesting that the items in figure 4-5 aren't important. But I am suggesting that you can easily defer consideration of them until later in the system development process. If you do, you'll be able to concentrate on those parts of the system that are critical to its success.

First, you can omit *extraction processes* from your model DFDs. Extraction processes draw information from existing data stores without making changes to those data stores. Although these are important processes in terms of providing information, you can leave them off your model DFDs. As long as the data structures and access paths you implement later on are sound, it's easy to add extraction processes to the system in a later phase of development.

Processes	Data flows	Data stores
Extraction File maintenance	Audit trail documents	Reference Archive

Figure 4-5   Components to omit from the model DFD

Second, you can omit *file-maintenance processes*. In any system, if you provide for making changes to data, you can be sure that, sooner or later, you'll have to provide for correcting it. Processes that do this are file-maintenance processes. The file-maintenance processes you must provide in your final system are determined by the structure of the database you use and the update policies you adopt. As a result, they too can be added to the system later on.

Third, you can omit *audit trail data flows*. Although a well designed system will produce an audit trail that documents the changes to the database, you don't need to include these data flows in a model DFD. When you design the processes that change databases, you'll know that you must provide audit trail data flows. Until that time, don't worry about them.

Fourth, you can omit data stores that represent collections of data that are designed to provide reference information. I call these *reference data stores*. In the brokerage system, for example, I had to include a reference data store that contains valid commodity codes. This data store isn't updated by the mainline processing programs of the system but is only referenced by them. Although you should recognize that you'll need reference data stores in your systems, you can leave them off your model DFDs.

Fifth, you can omit data stores that simply contain information that's stored away and not used again by central processes of the system. I call these *archive data stores*. For instance, the files of signed contract copies in the brokerage system are archive data stores. Someone might want to refer to a signed contract once in a while, but these data stores aren't used by any of the processes of the system. In an automated system, an archive data store might be a transaction log file that contains an entry for every transaction of a specific type processed in a given period.

If you obey this first principle for creating model DFDs, you'll be able to focus on the critical components of the system. Then, you can add the detail you've omitted at an appropriate later phase in the development process.

The second principle for creating model DFDs is: *Use data stores to connect all of the processes of the system.* If you obey this principle, you'll be forced to identify the data stores that are central to the system. Later on, these will become files or database views. Because the design of the file structures or database of a system is so critical in system development, I can't stress the importance of this principle enough. So I'll emphasize it as I show you how I developed the model DFD for the brokerage system.

You may be interested to know that we developed this principle as a solution to a couple of problems that had bothered us whenever we developed systems. Using other modelling methods, we would often discover late in the design process that we had failed to provide a required file. Or, we would discover that what we were treating as two files should logically be treated as one. Once we developed this principle, however, we eliminated oversights like this completely.

Developing the model DFD for the new brokerage system

Now that you've seen the general procedure for modelling a system, I'd like to show you how I developed my model DFD for the brokerage system. To a large extent, the set of DFDs in figures 4-6 through 4-10 speaks for itself. So I'll only try to give you the highlights of the development process.

Starting the model DFD When you create an analysis DFD, you start with the data flows on the context DFD for the old system. Similarly, you start with the data flows on the context DFD for the new system when you start a model DFD. Figure 4-6, then, shows my starting point for the model DFD. As you can see, the data flows correspond to those on the context DFD in figure 4-4. You can also see that this starting diagram isn't much different than the diagram with which I

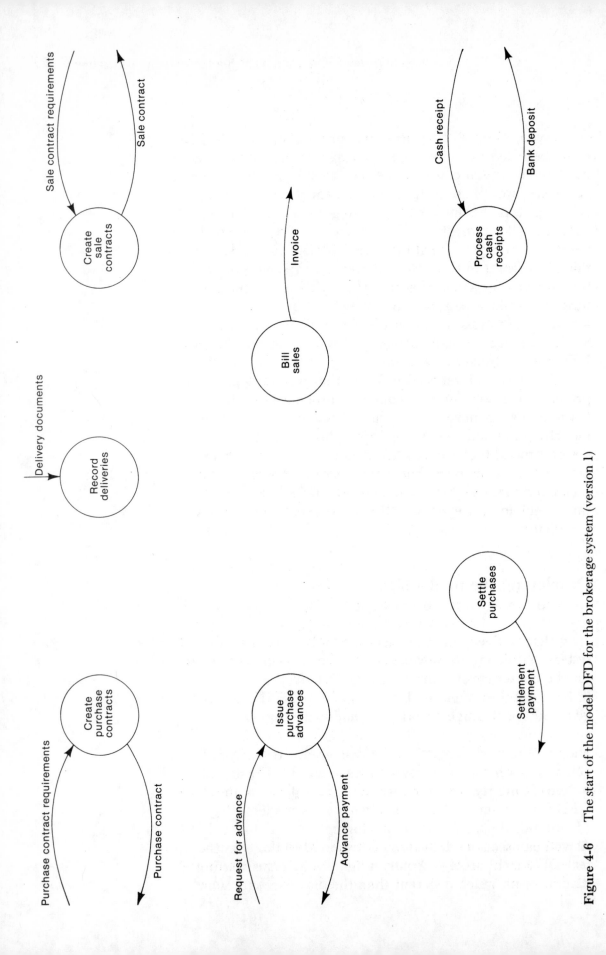

Figure 4-6   The start of the model DFD for the brokerage system (version 1)

began the development of the analysis DFD in the last
chapter. Here again, I've drawn a process at the end of each
data flow or group of related data flows.

Please notice that I grouped a pair of data flows that
weren't grouped when I did the analysis DFD: BANK
DEPOSIT and CASH RECEIPT. These are clearly related
since it's from cash receipts that the bank deposit is as-
sembled. I can group them now because I'm modelling the
new system. When I created the analysis DFD, I had to present them as they were in the old system.

Establishing the first connections   At this point, you're ready
to begin to use data stores to connect processes. As I see it, the
sensible place to start is with the processes that are central to
the system or that logically precede other processes. So figure
4-7 illustrates the first connections I made.

If you study figure 4-7, you'll see that all of the connections between the processes are through data stores. That is,
no data flows go directly from one process to another. This,
of course, is consistent with the second principle for creating
model DFDs. If I do this throughout the creation process, I'll
end up with a model that provides a solid basis for designing
an effective file structure or database.

Remember from the last chapter that a contract must be
prepared before any transactions related to it can be processed. Accordingly, my first addition to the diagram was a
pair of contract data stores connected to the contracting processes by data flows. Because contract information is required
to process deliveries, I next connected the two contract data
stores to the process RECORD DELIVERIES.

The processes BILL SALES and SETTLE PURCHASES
both require detailed information about deliveries. As a
result, I connected those processes to RECORD DELIVERIES
through another data store I called DELIVERIES. Also, contract information is needed at settlement and billing time, so I
connected the proper contract data stores with those processes
too.

Controlling advances to sellers   On the list in figure 4-1, one
of the functional requirements for the new system is to insure
that double payments not be made to sellers when advances

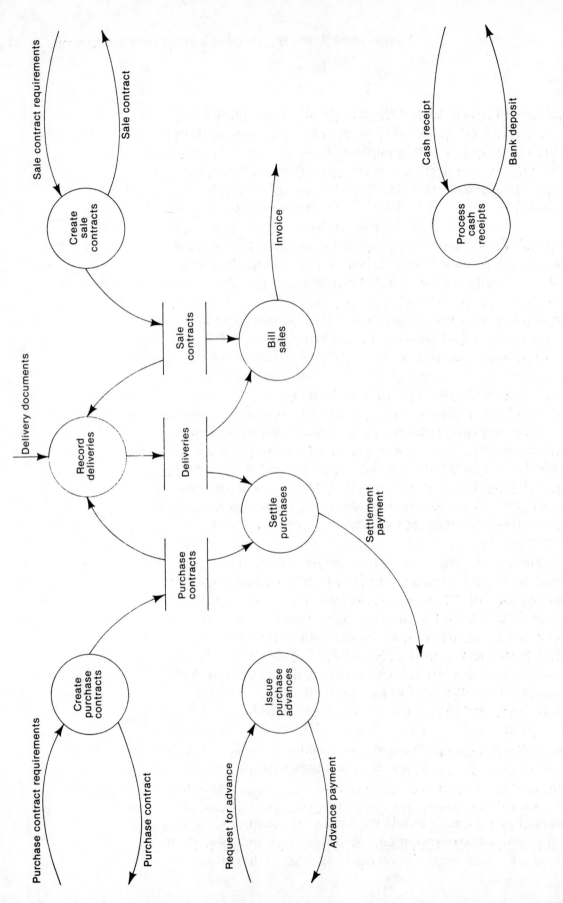

Figure 4-7 The model DFD for the brokerage system (version 2)

have already been issued. As a result, it's necessary for information about advances to be available to the process SETTLE PURCHASES. So I connected the processes ISSUE PURCHASE ADVANCES and SETTLE PURCHASES through a data store called PURCHASE ADVANCES. Figure 4-8 illustrates this connection.

Controlling receivables   Figure 4-8 also shows how I connected the processes that deal with collections. All I did was add a data store called RECEIVABLES and connect it to the processes logically related to it. If you compare this with its representation in the analysis DFD, you can see that the model for the new system is simpler than the way it was designed in the old system.

Controlling payables   Another requirement for the new system is that payments, whether for advances or for final settlements, be issued from the same process. To accommodate this requirement, I needed to get rid of the output data flows ADVANCE PAYMENT and SETTLEMENT PAYMENT and replace them with a single data flow coming from its own process. That new process had to be connected to the existing processes ISSUE PURCHASE ADVANCES and SETTLE PURCHASES.

Again, my goal is to connect processes through data stores, so I had to add an intermediate data store to make that connection. Figure 4-9 shows the next version of the model DFD. Here, I've added a data store I called PAYABLES into which flow the outputs of ISSUE PURCHASE ADVANCES and SETTLE PURCHASES. These are held until they're used by the new process, ISSUE CASH DISBURSEMENTS, to produce the new system output, CASH DISBURSEMENT.

Controlling advance billings   Just as the new system must not issue settlement payments when unapplied advances are outstanding, it must not issue final billings when unapplied advance billings are outstanding. To accommodate this requirement, I added a new data store to the sales side that's analogous to PURCHASE ADVANCES. It's called SALE

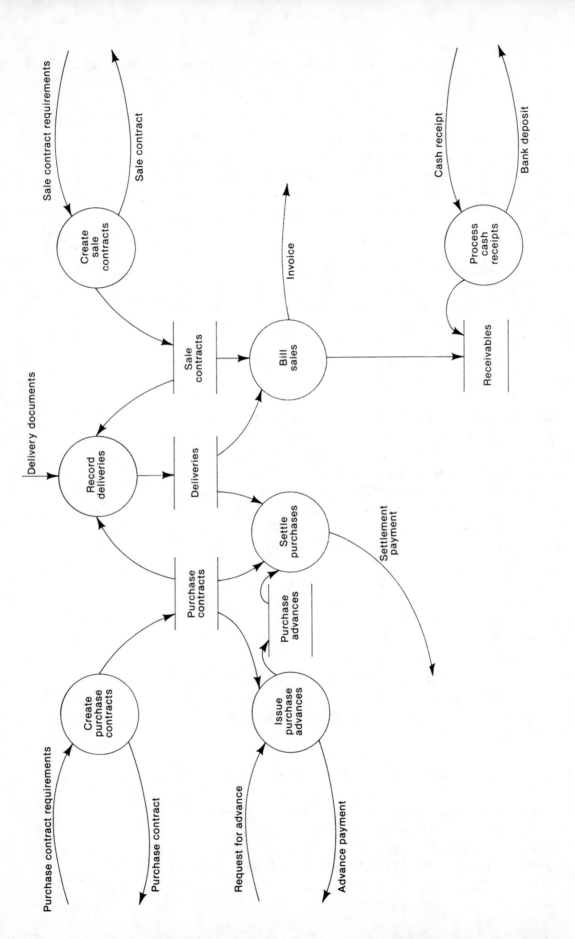

Figure 4-8 The model DFD for the brokerage system (version 3)

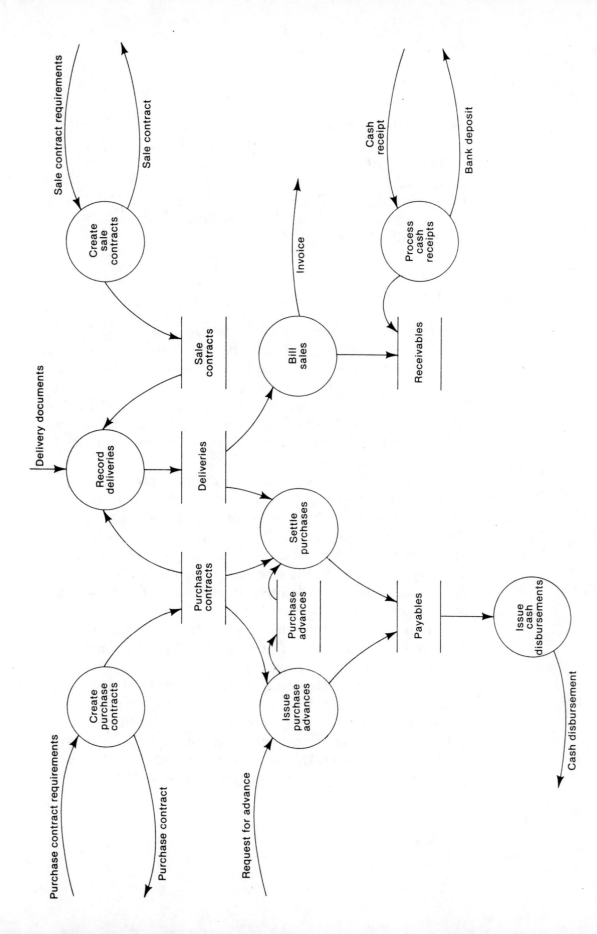

Figure 4-9 The model DFD for the brokerage system (version 4)

ADVANCES as shown in figure 4-10. I also added a new process called ISSUE ADVANCE BILLINGS. This process is connected to BILL SALES by the new data store. In the old system, all billings were done by the same process, which led to frequent double billings. So this new process and data store should solve the problem. With this done, the model DFD in figure 4-10 is complete.

Discussion

What I've created here is a model for the new system that shows the critical data interfaces between processes. I've done this by obeying the two principles for creating model DFDs. First, I've simplified the system as much as possible. Second, I've only connected processes by way of data stores. Certainly, my final system will be far more complex than this model DFD indicates. But this DFD is an excellent starting point for designing that final system.

In the stepping-stone approach I'm presenting in this book, the next step is to plan the system's data structures by developing in detail the contents of each of the data stores that I've shown in figure 4-10. The next chapter describes how I go about that. Then, after the system's data structures are planned, I expand the processes in figure 4-10 to reflect in detail what goes on inside each. In chapter 6, I'll explain that phase of system development.

If you compare this with other methods for modelling a new system, I think you'll find that this method is much simpler. And that, of course, is its strong point. By deferring detail until later development phases, you free yourself to concentrate on the critical processes and data flows of the system. By connecting all processes by way of data stores, you quickly identify the critical data stores of a system. After you see how the model DFD is used in subsequent design phases, I think you'll better appreciate the importance of this simple modelling procedure.

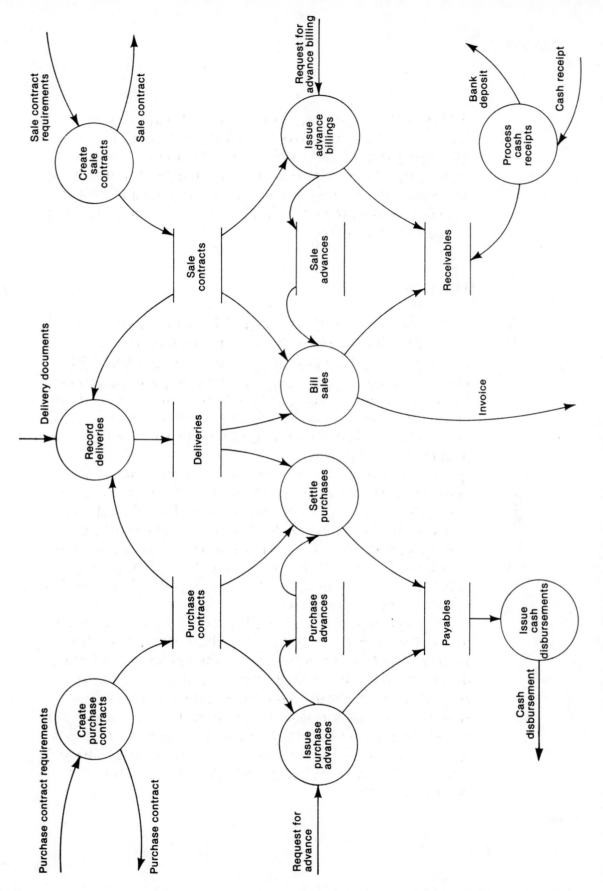

Figure 4-10 The complete model DFD for the brokerage system (version 5)

Terminology

extraction process
file-maintenance process
audit trail data flow
reference data store
archive data store

Objectives

1. List and describe the five types of system components you should *not* include in a model DFD.

2. Given an analysis DFD and a list of changes in the functional requirements for the new system, create a model DFD for the new system.

Part

Design

Chapter 5

How to define the data requirements of a system using data dictionary notation

 After you create the model DFD for a new system, you should start to define its data requirements. In this chapter, then, I'll show you a technique that will help you understand and organize the data components of a new system. As you start to define those data components, you'll be taking an important step toward database design as presented in chapter 7.

As you read this chapter, you should keep in mind that the four phases of system design (chapters 5 through 8) aren't done in a rigid sequence. Instead, you overlap these phases as described in chapter 1. After you model the new system, I recommend that you define the data requirements of the data stores on your model DFD. But that doesn't mean you'll have defined all of the data requirements of the new system. As you do the other phases of system design, you'll continue to expand and improve these preliminary definitions using the techniques of this chapter.

Data dictionary concepts

Using data dictionary terminology, the data flows and data stores of a system are data structures. A *data structure* is a combination of data elements that has a specific organization.

A *data element* in turn is an item of information that cannot be broken down into meaningful, subordinate components (or doesn't need to be). Frequently, a data structure will itself contain subordinate data structures.

A *data dictionary* is a comprehensive collection of information that fully documents the contents of every data flow and data store of a system. For each data structure and data element in a system, a data dictionary contains one entry. The entry for a data structure indicates what components it contains. The entry for a data element indicates its characteristics. A *data dictionary notation* is used within a data dictionary to show the relationships between the components of a data structure.

As you can imagine, developing and maintaining a data dictionary is a tedious, time-consuming task. When you're working on a massive project, however, it's often worth the time and money to set up and maintain a data dictionary. When many people are working on a large project, a data dictionary can reduce potential communication problems. In general, the more complicated a system becomes and the more developers working on it, the more necessary it becomes to adopt the systematic approach to database design that a data dictionary forces upon you.

If you do work on a project that requires a data dictionary, chances are you won't have much to say about how it's implemented or what notational system you'll use. Those decisions will have already been made for you. In such a case, it's likely that an automated data dictionary package will be in use. If a project is so large that it requires the development of a data dictionary, the data dictionary will probably be too large to maintain manually.

For more modest projects, however, I don't encourage you to develop a data dictionary. To create and maintain one is a costly undertaking. Moreover, when a data dictionary is implemented, it too often becomes an end in itself, an item of documentation that isn't used enough to warrant its high development and maintenance costs. So if it's likely that you wouldn't get much use out of the data dictionary you'd create, don't create one.

Instead, I recommend that you use data dictionary notation without implementing a data dictionary. In this case,

---

Address book entry
    Name
    Address
    City
    State
    Zip code
    Telephone number

Figure 5-1   A contents list for ADDRESS BOOK ENTRY

you make *contents lists* for the data flows and data stores of the system. These contents lists define the data requirements for the data structures of the system. As you create them, you should have one purpose in mind: to create documents that will help you define the new system's database more effectively later on. As you'll see, there's little value in developing a data dictionary for a system if you can design the database for the system in a more efficient way.

Developing your own data dictionary notation

Although data dictionary techniques are widely used, there's no standard for dictionary notation. As a result, you should use the notational system that's the simplest and easiest for you to understand when you create contents lists. In the next few pages, then, I'll first explain what a notational system needs to provide. Then, I'll illustrate the notation I use, along with two others. After you review these methods, you can develop your own notation by selecting the parts you feel are most efficient.

Using a notational system   As a simple example, consider the information you record when you make an entry in your personal address book. The entry is a data structure because it consists of data elements arranged in a specific way. For most of us, a typical entry would consist of a name, address, city, state, zip code, and telephone number. In figure 5-1, you can see a contents list for a data structure with those components.

Address book entry

    Name

    Address

    City

    State

    Zip code

    Telephone number

        Area code

        Local telephone number

Figure 5-2   A contents list for ADDRESS BOOK ENTRY that shows a second level of subordination

In this list, indentation shows the subordination of data elements within a structure. And subordination is something that any notation system must be able to show.

When you organize the components of a data structure, you'll find that subordination can extend down many levels. Although the subordinate elements in figure 5-1 extend down only one level, it would be reasonable to show that TELEPHONE NUMBER consists of an area code and a local (seven-digit) number. If you wanted to show this additional level of subordination, you could do so as shown in figure 5-2.

When you decide how far down to expand a data structure, you need to use common sense. For example, you might know that the ADDRESS component of ADDRESS BOOK ENTRY also consists of these components: street number, street name, and apartment number. Whether or not you choose to reflect that lower-level structure in the contents list depends on how important the extra detail is. If the extra detail is irrelevant, omit it.

In addition to showing subordination, a notational system must also be able to represent other characteristics of data structures in a contents list. First, the notation must be able to show whether a component is required or optional. Second, it must be able to show how many times a structure or element is repeated. And finally, it must be able to show that one item must be selected from a group of two or more. In the next section, I'll show you a data dictionary notation you can use to indicate these characteristics.

From the point of view of physical database design, it may not matter whether some items are optional or a selection from three elements must be made. But from the point of view of process design, those are important considerations. If, for example, you make a record description out of the address book entry, your description needs to provide for all of the data elements that make up the structure, whether or not they're optional or selections. In terms of database design, then, these characteristics are irrelevant. On the other hand, these factors do affect the processes that fill in the record with data. So contents lists aren't useful just when you design a database; they're also useful when you design processes.

My notational system   When I create contents lists, I use a form I designed for the purpose. I fill out one page of the form for each data structure I choose to define (more on how I decide which data structures to define in a moment). Don't feel compelled to adopt a special form, however. You may feel more comfortable using a simple legal pad.

Figure 5-3 shows two contents lists recorded on my form. These lists enhance and replace the ADDRESS BOOK ENTRY you've already seen. As simple as they are, they illustrate everything you need to know about my notational system. At the top of the form is a space for the name of the data structure being defined. Then, the form is split into four columns. The widest of the four is for the names of the structure's subordinate components. The rightmost column is for comments about the data components listed. The two columns on the left labelled "group" and "repetitions" provide additional information about the components of the structure.

When you fill out this form, you can show in-line subordination of data elements within a data structure just as I did in figure 5-2. In fact, if you compare figures 5-2 and 5-3, you'll see I represented the data structure TELEPHONE NUMBER the same way in both. That is, I used in-line subordination.

As an alternative, though, you can show the contents of a subordinate data structure in a separate list. That's how I represented the data structure INDIVIDUAL NAME in figure 5-3. When you make a separate list for a subordinate data structure, you indicate it by putting a "Y" in the "group" col-

Data structure: Address book entry

Group	Repetitions	Components	Comments

Y	0-2	{ Individual name / Company name }
		{ Street address / Post office box }
		City
		State
		Zip code
		Telephone number
		  Area code
		  Local telephone number	

Data structure: Individual name

Group	Repetitions	Components	Comments
	0-1	First name Middle initial Last name	

Figure 5-3 A contents list for ADDRESS BOOK ENTRY using my form and my data dictionary notation

umn for that item in the higher-level list. If you do this, another reader can see at a glance which items in a data structure are expanded by lower-level lists. Also, you'll be reminded when you use the lists later that the items are data structures. Here, I entered a "Y" in the group column for INDIVIDUAL NAME in the list for the higher-level data structure ADDRESS BOOK ENTRY.

As a general guideline, if a subordinate data structure occurs in only one higher-level data structure, expand the data structure using in-line subordination. If you bother to make a separate contents list for such a data structure, you'll have to do some unnecessary paging when you use the lists.

On the other hand, if a data structure is a component of two or more higher-level structures, make a separate list for it. This method has two advantages over showing subordination in-line. First, it reduces the size of your contents lists by eliminating duplicated entries. Second, if a structure must be changed, the change only needs to be made in one place. In contrast, making the change in each place the in-line subordination appears can be time-consuming and error-prone. Of course, the disadvantage to making a separate list for a low-level structure is that the user of the lists must turn back and forth between pages to get a complete picture of an entire data structure. In my opinion, though, this disadvantage is outweighed by the advantages.

Although other notational systems use separate symbols to indicate optional and repeated items, I prefer to keep my use of symbols to a minimum. This avoids the confusion that combinations of symbols can cause. As a result, instead of using symbols to show an optional or repeated item, I enter the item's range of occurrences in the column labelled "repetitions" on my form. If an item has a range starting with zero, it means it's optional. In figure 5-3, for example, the data element MIDDLE INITIAL is optional in INDIVIDUAL NAME. To show that, I entered 0-1 in the "repetitions" column for that data element. Also, in this enhanced version of the ADDRESS BOOK ENTRY, from zero to two occurrences of TELEPHONE NUMBER are allowed.

Although I avoid symbols to indicate optional and repeating items, I do use braces to indicate a selection. This shouldn't be confusing, however, because I don't use braces in combinations with other symbols. In figure 5-3, I've indicated two selections. The first selection is from INDIVIDUAL NAME and COMPANY NAME; the second is from STREET ADDRESS and POST OFFICE BOX.

To avoid confusion, don't include any in-line subordination within the braces that indicate selection. If any of the items to be selected are data structures, expand those data structures on separate forms whether or not they occur only within the higher-level data structure. In figure 5-3, I used the "group" column of the form to show that INDIVIDUAL NAME is a data structure expanded in another list.

Data structure: Address book entry

Group	Repetitions	Components	Comments
	0-1	⎧ Individual name ⎫ ⎪ First name ⎪ ⎨ Middle initial ⎬ ⎪ Last name ⎪ ⎩ Company name ⎭ ⎰ Street address ⎱ ⎱ Post office box ⎰ City State Zip code	

	0-2	Telephone number
		  Area code
		  Local telephone number	

Figure 5-4   An alternative contents list for ADDRESS BOOK ENTRY using my notation

Because this is a simple example, expanding
INDIVIDUAL NAME in-line as shown in figure 5-4 probably
wouldn't confuse anybody. So it's up to you to decide whether
to expand a selection item in-line or on a separate form.
Remember, though, that your goal should be to make the
contents lists as easy as possible to use.

Alternative notational systems   Since contents lists are only
intended to be working tools, the notational system you use
isn't that critical. As long as you're comfortable with the
notation you use and it helps you work more effectively, it
doesn't make that much difference. As a result, I'll now present two other notational systems. Both are illustrated using
the ADDRESS BOOK ENTRY as an example, so you can
compare these systems with mine.
    Using the first method as shown in figure 5-5, a separate
list (actually, a separate data dictionary entry) is required for

Address book entry = $\begin{bmatrix} \text{Individual name} \\ \text{Company name} \end{bmatrix}$ +

$\begin{bmatrix} \text{Street address} \\ \text{Post office box} \end{bmatrix}$ +

City +

State +

Zip code +

$_0^2${Telephone number}

Individual name = First name +

(Middle initial) +

Last name

Telephone number = Area code +

Local telephone number

Note: A complete implementation of this notational system would require a separate data dictionary entry for each data element.

Figure 5-5   A contents list for ADDRESS BOOK ENTRY using a second notational system

each data element and each data structure. To relate the components of a data structure, the plus sign (+) is used. Braces, preceded by numbers representing a range, indicate repeated items. Parentheses indicate optional items. Brackets indicate selections. And no effort is made to show subordination within the data structure by indentation. Instead, the user must refer to a separate entry for a lower-level component to see if it's a data element or a data structure. If it's a data structure, the user needs to follow each of its components to their entries also.

As I see it, this method requires too much unnecessary work because it requires a separate entry for each data element. So this may be a case in which the method became more important than the result. The notation itself, however, has a certain appeal to it.

Figure 5-6 illustrates a second notational system. This one is closer to mine, but it too (if fully illustrated) requires a

---

Address book entry

$\left\{\begin{array}{l}\text{Individual name}\\ \text{Company name}\end{array}\right\}$

$\left\{\begin{array}{l}\text{Street address}\\ \text{Post office box}\end{array}\right\}$

City

State

Zip code

Telephone number * (0-2)

Individual name

First name

[Middle initial]

Last name

Telephone number

Area code

Local telephone number

Note: A complete implementation of this notational system would require a separate data dictionary entry for each data element.

Figure 5-6   A contents list for ADDRESS BOOK ENTRY using a third notational system

separate entry for each data element. It also relies heavily on symbols. Here, braces indicate selection, and brackets indicate an optional item. Repeated items are followed by an asterisk and the range of repetition in parentheses.

Figure 5-7 summarizes my notation and the notation of the two alternative systems I just presented. It also gives the notation used in COBOL manuals to show the syntax of that language. Since many system developers come from COBOL programming backgrounds, they may find this notation the most natural of all. As you can see, not only are the notational systems inconsistent, but in some cases they're actually contradictory.

	My method	1st alternative	2nd alternative	COBOL programming language manuals
In-line subordinate data structures	indicated by indentation	not allowed	indicated by indentation	not applicable
Optional item	(0-1) indicated in repetitions column for item	parentheses around item name (item)	brackets around item name [item]	brackets around item name [item]
Repeated item (repeated from a to b times)	(a-b) indicated in repetitions column for item	braces around item name with range preceding leading brace $^{b}_{a}\{$item$\}$	item name followed by asterisk and range in parentheses item *(a-b)	indicated by ellipses without range item ...
Selection of one of two or more items	braces around item names $\{$item-1$\}$ $\{$item-2$\}$	brackets around item names [item-1] [item-2]	braces around item names $\{$item-1$\}$ $\{$item-2$\}$	braces around item names $\{$item-1$\}$ $\{$item-2$\}$

Figure 5-7   A summary of the features of four notational systems

Applying data dictionary notation

Now that you're familiar with data dictionary notation, I'd like to show you how to apply it to the design process in a sensible way. Your goal at this point is not to begin file or database design. It's simply to develop preliminary contents lists for the data stores on your model DFD. As you develop the contents lists, you'll identify the data components of each data store.

To specify a data store's contents, you have to start with the outputs that contain information drawn from it. Fundamentally, a data processing system exists only to give users output for one purpose or another. So a system's data stores don't exist for their own sakes. Instead, their purpose is to hold information that's processed to create useful output. In database design, then, you must make sure that each data

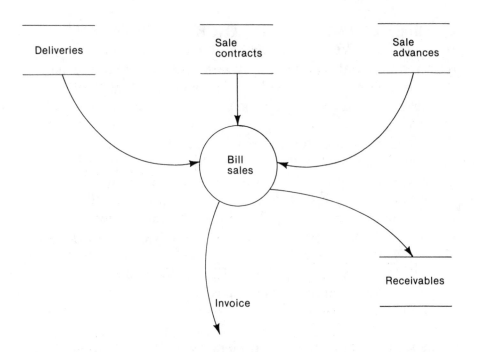

Figure 5-8   A portion of the model DFD for the brokerage system

store contains the information its related processes require to produce their outputs. And you start to do this by developing contents lists.

To develop a preliminary contents list for each data store, work from the model DFD. In other words, use the output data flows on the model DFD to determine what the data stores must contain. Because some output items were left off the model DFD, you can be sure that it will be necessary later to add to these preliminary contents lists. But since you need to start somewhere to determine the contents of the data stores, the model DFD is the sensible place to begin.

The first step in developing the contents lists for the critical data stores, then, is to create contents lists for each of the output data flows. Then, you can work from these lists to determine what the related data stores must contain. To illustrate this process, I'm going to show you how I developed the contents lists for just a few of the data stores in the brokerage system, the ones shown on the portion of the model DFD in figure 5-8. But before I can develop the contents lists for the data stores, I must develop the contents list for INVOICE.

Creating a contents list for INVOICE To develop the contents list for INVOICE, I started by trying to determine the data requirements of the new system's invoice. I started with the manually prepared invoice and concluded that these groups of components would be required on the new invoice:

identification of the buyer (name and address)
identification of the related contract (number)
identification of the invoice (number and date)
description of the billing (commodity and quantity)
financial totals (price, charges, premium, discount,
    reductions for previously issued advance billings, and
    final receivable total)
substantiation of the financial totals (delivery and
    advance application details)
remarks

My next step was to identify in detail what each of these groups consists of.

Although most of these groups of components seemed straightforward to me, I had my doubts about the group called "financial totals." When you're faced with a group of related data elements that you're not sure of, I suggest that you list them, their relationships, and their sources in some sort of table. Just by forcing yourself to look at them in a systematic way, you'll be taking a large step toward understanding them. Figure 5-9, then, is the table I prepared to help me understand exactly what "financial totals" should contain.

If you argued that the information in figure 5-9 is really closer to process specification than database design, I wouldn't disagree. After all, information about *how* data items are manipulated is clearly procedural. However, as I've pointed out, process specification and database design overlap. To understand the data elements that make up the group "financial totals," I had to look closely at the way they're derived.

If you look carefully at figure 5-9, you'll see that most of the data elements are calculated from, or are direct pick-ups from, the data elements in the related data stores. However, two of the data elements, UPWARD ADJUSTMENT

Data elements and relationships	Sources
Total quantity billed	Calculated at time of billing from weights of all items selected from data store DELIVERIES
× Sale contract price	SALE CONTRACTS
Invoice extension	Calculated at time of billing from other INVOICE data elements
– Total of charges on deliveries	Calculated at time of billing from charges on all items selected from data store DELIVERIES
Invoice subtotal	Calculated at time of billing from other INVOICE data elements
+ Upward adjustment (premium)	Supplied at time of billing by operator (not from a data store)
– Downward adjustment (discount)	Supplied at time of billing by operator (not from a data store)
Invoice total	Calculated at time of billing from other INVOICE data elements
– Advance billing total deducted from invoice total	Calculated at time of billing from unapplied advances on related items in data store SALE ADVANCES
Final receivable total	Calculated at time of billing from other INVOICE data elements

Figure 5-9   A table showing the sources of the financial data elements of INVOICE

(PREMIUM) and DOWNWARD ADJUSTMENT (DISCOUNT),
come from another source altogether: operator entry.
But since I'm only concerned now with data store
specification, I can consider those data elements later on
when I begin to design the process for BILL SALES. So here
again you can see how database design and process specification are intertwined.

At any rate, after I gained an adequate understanding of
the "financial totals" group, it was easy for me to develop the
contents lists for INVOICE as presented in the three parts of
figure 5-10. Although this set of lists is complicated, you
should understand my notation well enough to figure out
what this set of lists represents. As you can see, the lists are
arranged in alphabetical order. If you look at my entry in the
"repetitions" column for INVOICE DELIVERY DETAIL in
part 2 of figure 5-10, you can see it indicates that the structure can occur one or more times with no upper limit.

Data structure: Advance billing deduction detail

Group	Repetitions	Components	Comments
		Advance invoice number	
		Amount applied to final invoice from advance invoice	

Data structure: Buyer name and address

Group	Repetitions	Components	Comments
Y		{ Individual name / Company name }	
		Address	
		City	
		State	
		Zip code	

Data structure: Delivery charge detail

Group	Repetitions	Components	Comments
		Charge description	
		Charge base amount	
		Charge extension	Base amt × deliv net wt

Data structure: Individual name

Group	Repetitions	Components	Comments
	0-1	First name	
		Middle initial	
		Last name	

Figure 5-10 Contents lists for the INVOICE data flow and its subordinate data structures (part 1 of 3)

Data structure: Invoice

Group	Repetitions	Components	Comments
		Invoice number	
		Invoice date	
		Invoice month	
		Invoice day	
		Invoice year	
		Sale contract number	
Y		Buyer name and address	
		Commodity	
		Total quantity billed	
		Sale contract price	
		Invoice extension	
		Total of charges on deliveries	
	0-1	Upward adjustment (premium) data	
		Upward adjustment (premium) amount	
		Upward adjustment (premium) description	
	0-1	Downward adjustment (discount) data	
		Downward adjustment (discount) amount	
		Downward adjustment (discount) description	
		Invoice total	
		Advance billing total deducted from invoice total	
Y	0-10	Advance billing deduction detail	
		Final receivable amount	
Y	1 or more	Invoice delivery detail	
	0-5	Invoice remark	

Figure 5-10   Contents lists for the INVOICE data flow and its subordinate data structures (part 2 of 3)

Data structure: Invoice delivery detail

Group	Repetitions	Components	Comments
		Carrier	
		Carrier bill of lading number	
		Weight certificate	
	0-1	Grade certificate	
		Origin location	
		Delivery date	
		Delivery month	
		Delivery day	
		Delivery year	
	0-1	Delivery remark	
	0-1	Delivery freight charge	
	0-1	Delivery dockage percentage	% Garbage in delivery
		Delivery gross weight	
		Delivery tare weight	Weight of truck or RR car
		Delivery net weight	
Y	0-10	Delivery charge detail	
		Delivery net price	

Figure 5-10   Contents lists for the INVOICE data flow and its subordinate data structures (part 3 of 3)

Creating preliminary contents lists for data stores   Before I could develop the contents lists for the data stores shown in the DFD in figure 5-8, I had to determine the sources of all of the data elements that make up INVOICE. In the table in figure 5-9, I had already identified the sources of the data elements of "financial totals." To identify the sources of the rest of the elements, I used common sense and my understanding of the firm's operations.

As you can see in figure 5-11, I used the "comments" column of the INVOICE contents list to record the source of each data item. You'll notice here that I didn't bother to list

Data structure: Invoice

Group	Repetitions	Components	Comments
		Invoice number	Control item
		Invoice date	Control item
		Invoice month	
		Invoice day	
		Invoice year	
		Sale contract number	Sale contracts data store
Y		Buyer name and address	Sale contracts data store
		Commodity	Sale contracts data store
		Total quantity billed	Calculated (figure 5-9)
		Sale contract price	Sale contracts data store
		Invoice extension	Calculated
		Total of charges on deliveries	Calculated (figure 5-9)
	0-1	Upward adjustment (premium) data	Operator entry
		Upward adjustment (premium) amount	
		Upward adjustment (premium) description	
	0-1	Downward adjustment (discount) data	Operator entry
		Downward adjustment (discount) amount	
		Downward adjustment (discount) description	
		Invoice total	Calculated (figure 5-9)
		Advance billing total deducted from invoice total	Calculated (figure 5-9)
Y	0-10	Advance billing deduction detail	Sale advances /calculated
		Final receivable amount	Calculated (figure 5-9)
Y	Any #	Invoice delivery detail	Deliveries
	0-5	Invoice remark	Operator entry

Figure 5-11   A contents list for INVOICE showing the sources of its components

in detail the sources of all of the data elements from "financial totals" since they're identified in the table. After all, both figures are just working papers, not final documentation. Their purpose is to make it easier for me to create preliminary contents lists for the three data stores from which INVOICE's components are drawn. As long as I feel comfortable that I've identified the sources of all of the data items in the structure, I don't have to formalize my working papers.

As you can see in figure 5-11, INVOICE's data elements not only come from three data stores, but some come from operator entry. Still others are in another class altogether called *control items*. Control items are supplied by a system to identify transactions and to provide standard information. Here, INVOICE NUMBER will be an identifying number assigned in sequence to each invoice as it's created. INVOICE DATE will be drawn from the system as the current date. Both of these, then, are control items.

Because I'm only trying to identify the data elements that make up the data stores on the model DFD at this point, the details of how control items are stored and accessed aren't important. They may be supplied by a computer's operating system (such as date and time), maintained in a common work area for access by many users, or stored in private data stores. Sometimes control items are used as keys for accessing data structures. As such, they're an important factor in database organization. But I still don't have to concern myself with them when I develop contents lists.

After I identified the sources of each of the components of INVOICE, it was an easy matter to enter them in preliminary contents lists for the three input data stores of figure 5-8. At this point, the contents of the three related data stores that are required for INVOICE are self-evident. As a result, I'll only show you the one for the data store SALE CONTRACTS. It's in figure 5-12.

Of course, to complete the contents lists for the system's critical data stores, I had to examine all of the output data flows on the model DFD. Whenever I identified a new data element derived from a data store on the diagram, I added it to that data store's preliminary contents list. I could illustrate the complete development of each data store's preliminary contents list, but I think you've seen enough now to under-

Data structure: Sale contracts

Group	Repetitions	Components	Comments
Y		Sale contract number Buyer name and address Commodity Sale contract price	

Figure 5-12 A preliminary contents list for the SALE CONTRACTS data store after creating the contents list for INVOICE

stand the process. Instead, I'll just show you the final list for the SALE CONTRACTS data store. It's shown in figure 5-13. You can see that it includes the four data items required by BILL SALES (the ones I listed in figure 5-12) plus several others. I identified the other components by determining the data requirements for the other processes that draw from SALE CONTRACTS (RECORD DELIVERIES and ISSUE ADVANCE BILLINGS).

Three guidelines for creating contents lists The examples in figures 5-10 and 5-13 illustrate three points you should keep in mind when you create preliminary contents lists. First, you can't implement a repeating structure with an unlimited or large number of occurrences, like INVOICE DELIVERY DETAIL, as part of another data structure. The only solution is to make each repetition a separate occurrence of the entire structure for the data store. In this case, if a given buyer's invoice contains five INVOICE DELIVERY DETAIL groups, that means five occurrences of the structure for the DELIVERIES data store. When I discuss physical database design in chapter 7, I'll have more to say about repeating groups.

Second, when you identify a data element in an output data flow that's not associated with a *single* occurrence of the data store structure you're interested in, you shouldn't include it in the contents of the data store. In the INVOICE example, the data element TOTAL QUANTITY BILLED is calculated

Data structure: Sale contracts

Group	Repetitions	Components	Comments
Y		Sale contract number	
		Buyer name and address	
		Commodity	
		Sale contract price	
		Open/closed status	
		Commodity specification	
		Commodity variety	
		Commodity quality	
		Commodity comment	
		Sale contract quantity	
		Quantity already delivered on sale contract	
		Shipment terms	
		Shipment destination	
		Shipment method	
		Shipment date	
		Amount of dockage allowable	

Figure 5-13 A preliminary contents list for the SALE CONTRACTS data store after creating contents lists for all output data flows

based on the weight values for all deliveries invoiced. Clearly, then, it can't be associated with a single occurrence of the data store DELIVERIES. Equally obvious, however, is the fact that DELIVERIES must contain the original weight values from which the total is accumulated.

Third, it may not be necessary for a data structure with commonly used values to be stored in its entirety in a data store. For example, the data store SALE CONTRACTS contains the data structure BUYER NAME AND ADDRESS. But in the final implementation of the SALE CONTRACTS data store, that may not be the case. Instead, SALE CONTRACTS may contain some mechanism such as a key or pointer through which BUYER NAME AND ADDRESS can be retrieved from another data structure. At this point, though, you can feel comfortable if you show that the data store contains the complete data structure. I'll have more to say about this important point of database design in chapter 7.

Discussion

After you've created the preliminary contents lists for the critical data stores of a system, you're still a long way from designing the new system's database. Before you can do so, you'll need to add other data items to the lists as you look more deeply into the system's processes. Then, you'll have to plan physical implementations for all data stores to integrate them into the system's final database. At this point, you should only be sure that your contents lists contain all of the data elements necessary to allow the processes of the system to create the output data flows of the model DFD. And that's not a bad first step.

To give you some perspective about this phase of system development, you should only spend a few days creating contents lists after you create the model DFD for a system. When I developed the brokerage system, for example, I spent less than a week creating the preliminary contents lists. Also, since creating the contents lists is tedious work, you don't have to do it eight hours a day. You can work on contents lists for a few hours, start to organize the functions of the system for a few hours, go back to your contents lists, and so on, until you're done with them. In truth, I probably enjoy preparing contents lists less than any other phase of system development. But the contents lists justify their development time as you start to design the functions and database of the new system.

In the next chapter, I'll switch from data definition to process design. Then, after I've explained process design, I'll show you how to design a system's database. Specifically, I'll show you what you need to know to move from the preliminary contents lists for data stores to a physical database design.

Terminology

data structure
data element
data dictionary
data dictionary notation
contents list
control item

Objective

Working from a model DFD, use data dictionary notation to create preliminary contents lists for the data stores.

Chapter 6

How to organize the functions of a system using a system structure chart

 In chapter 4, I showed you how to identify the critical processes and data stores of a system by using a model DFD. The model DFD also shows the relationships between those processes and data stores. But once you've established those relationships, you need to examine and organize all of the system's processes. This isn't limited to the processes on the model DFD. You also need to include the processes you purposefully excluded from the model DFD to simplify it.

In this chapter, then, I'm going to show you how to use a system structure chart to organize all of the functions of a system. I'll show you why this chart is ideal for designing a menu-driven system. And I'll show you how you can use one to involve the user, plan a development sequence for the programs of a system, and monitor the progress of a system's development. But first let me explain why a DFD, although effective as an analysis document, isn't effective when it comes to designing the detailed functions of a system.

Why a DFD isn't effective
as a detailed design document

During analysis, the DFD is a uniquely valuable tool. By using it, I was able to identify the brokerage system's critical data stores. That's an important accomplishment. In fact, coordinating database design and process specification is perhaps the most difficult task in a system development project. And drawing the model DFD helps you identify the data stores through which the system's processes will relate to one another.

Unfortunately, however, using the DFD as a design document after you model a new system becomes overly complicated. In a typical system, the critical data stores are connected to many lower-level processes. As a result, a complete lower-level DFD that shows all of these connections becomes a spiderweb, both difficult to draw and confusing to use. In short, expanding the processes of a model DFD takes more effort than its benefits warrant.

As a simple example of this spiderweb effect, consider figure 6-1. This is an expansion of CREATE PURCHASE CONTRACTS from the brokerage system's model DFD (figure 4-10). As you can see, what the diagram indicates is that CREATE PURCHASE CONTRACTS really consists of four subordinate processes. For now, don't worry about what the four subordinate processes do or how I went about identifying them. I'll describe that in a moment. Just notice that each subordinate process has the same inputs and the same outputs. Although I couldn't draw this diagram until I identified the four subordinate processes, there was little point in drawing the diagram once I did. It doesn't show any relationships that aren't obvious, so it doesn't contribute to my understanding of the system. In fact, it's an unnecessarily complicated presentation of a simple concept.

Now I'm not trying to say that the DFD isn't of any use in design. In some cases, you'll want to use the DFD to clarify obscure or complex processes. Usually, however, you won't need to do so. Instead, you can proceed by working with another design tool that's more useful and easier to draw and maintain: the system structure chart.

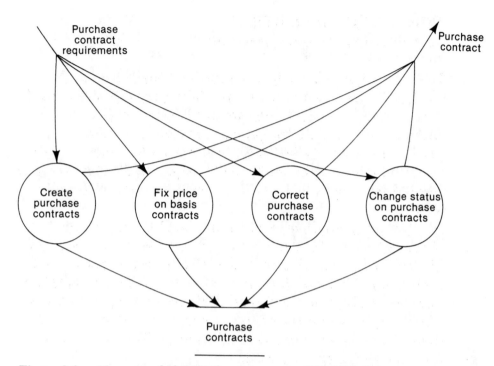

Figure 6-1   The expanded DFD for the process CREATE PURCHASE
CONTRACTS showing the complexity of the data flows in a
relatively simple expansion of a critical process

Why the system structure chart is effective as a design document

A *system structure chart* is a hierarchy diagram that shows
each of a system's processes. It doesn't show the relationships
of those processes through data stores as the DFD does. Instead, it shows the control relationships among the system's
processes (alternatively called *functions* or *modules*) by indicating levels of subordination.

The system structure chart looks much like the organization chart of a business. At the top of a business's organization
chart is a single box that represents the firm's top-level
manager. That top-level manager is responsible for all of the
activities performed by all of the individuals in subordinate
positions on the chart. Similarly, at the top of a system structure chart is a single box that represents the top-level system
function. Subordinate to that box, all of the functions that

comprise the business's operations are performed. Each of those functions, in turn, may be composed of subordinate functions at still lower levels. The top-level and mid-level functions serve only to control the functions beneath them in the hierarchy. The bottom-level functions are the real "workers" of the system. Typically, the higher a function is in the structure chart, the more general is its function.

By the time you finish this chapter, I think you'll see why the structure chart is such an effective design document. First, it shows all of the functions of a system in a simple graphic presentation. Second, it shows the relationships of all the functions in terms of which modules control which functions. And third, it's an excellent document for planning and monitoring the implementation of an entire system. In my experience, I've never seen another design document as effective for these purposes as the system structure chart presented in this chapter.

How to create a system structure chart

I follow a four-step process to develop a system structure chart. First, I create a preliminary chart from the model DFD for the system. Second, I expand each of the processes on the preliminary chart to show their subordinate functions. Third, I add processes to the chart that I omitted from the model DFD. Fourth, I expand these processes.

Step 1: Create the preliminary chart This is a simple, mechanical step. To create the preliminary chart, you start with the model DFD for the system. For the brokerage system, that's figure 4-10. I worked from it to create the structure chart in figure 6-2. Notice the one-to-one relationship between the processes on the model DFD and the second-level modules on the system structure chart. Also, notice that I gave each second-level module the same name as its corresponding DFD processs.

The top-level function on the chart is created simply to control the execution of the second-level modules. Conceptually, the top-level module represents the entire system. You can see in figure 6-2 that I named it using a description that reflects its broad, controlling function.

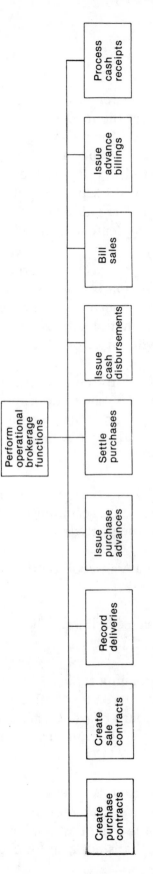

Figure 6-2   The system structure chart for the brokerage system based on the model DFD in figure 4-10

Step 2: Expand the original processes   After you've transferred the system's major processes from the model DFD to the system structure chart, you need to expand each of them. In this step, you have to consider what goes on in each process. This is the most challenging part of developing the system structure chart. As a result, you can expect to put some real creative effort into this step. The guidelines that follow should help, but you need to supplement them with your own analytical skills and insights.

When you drew the model DFD, you gave each process a name that indicates what its basic function is. In almost all cases, though, a structure chart function derived from a DFD process really consists of several subordinate modules. Of course, those subordinate functions will vary from application to application.

In the most general terms, you need to find out how each second-level process accomplishes what its name implies. Then, you need to determine how that transaction can be corrected or adjusted if necessary. Specifically, you should answer these questions for each module:

1.   What starts the module's basic task? In particular, must any preliminary or preparatory tasks be accomplished before the module's basic task can be started?

2.   What adjustments are allowed to a transaction after it has been processed?

3.   Can a transaction be deleted or cancelled altogether after it has been processed?

I find it useful to answer these questions in a table like the one in figure 6-3. This is the table I prepared for the functions of the brokerage system's preliminary chart. I listed each process name across the top of the page. Then, in three rows, I answered my three questions.

First, look across the row labelled "Started by?" For each of the second-level modules in figure 6-2, I reviewed what causes them to be performed. Notice that most of the transactions are started as a result of a user request or the receipt of some documents. However, that's not true for three

	Create purchase contracts	Create sale contracts	Record deliveries	Issue purchase advances	Settle purchases	Issue cash disbursements	Bill sales	Issue advance billings	Process cash receipts
Started by?	Broker request	Broker request	Receipt of delivery documents	Seller request	Selection from list of unsettled deliveries	Selection from list of unpaid payables	Selection from list of unbilled deliveries	Broker request	Receipt of check
Adjustments allowed?	Price fix Correction Status change	Price fix Correction Status change	Correction	Application adjustment	Not allowed directly (reverse with new opposite entry)	Discount Balance transfer	Not allowed directly (reverse with new opposite entry)	Application adjustment	Discount Balance transfer
Deletion allowed?	Not allowed	Not allowed	Allowed	Not allowed	Not allowed	Allowed— void check	Not allowed	Not allowed	Allowed— returned check

Figure 6-3 A table for analyzing the brokerage system's required functions

modules: SETTLE PURCHASES, ISSUE CASH DISBURSEMENTS, and BILL SALES. In each of these, transactions are processed after being selected from lists of items available for processing. Therefore, prior to performing the basic task in each of these modules, a preliminary list-preparation function must be performed. Obviously, there are other ways to implement these functions, but lists are what the user wanted.

For example, the basic task of SETTLE PURCHASES is started after the user reviews a list of all unsettled deliveries and selects the ones to be settled. To implement this, I'll need to add a list-preparation function to the system structure chart subordinate to SETTLE PURCHASES. In addition, I'll need to add similar modules subordinate to ISSUE CASH DISBURSEMENTS and BILL SALES.

Second, look at the last row, labelled "Deletion allowed?" Here, you see that most of the transactions performed by the system may not be deleted or cancelled after they're processed. In some cases, this is an audit trail consideration, while in others it's intended to insure the integrity of the system's database. (This second factor presupposes knowledge about how the system's database will be organized and illustrates that system design is an integrated process, not a series of unrelated, discrete steps.)

In figure 6-3, you can see that transactions from three modules (RECORD DELIVERIES, ISSUE CASH DISBURSEMENTS, and PROCESS CASH RECEIPTS) may be deleted. As a result, I'll have to include three deletion functions on my expanded system structure chart.

Finally, look at the middle row of figure 6-3, labelled "Adjustments allowed?" The entries in this row require some explanation. For each of the contracting modules (CREATE PURCHASE CONTRACTS and CREATE SALE CON-TRACTS), three kinds of adjustments are allowed. A "price fix" is a special kind of adjustment for contracts that were created without a definite fixed price (these are called "basis" contracts). A "status change" will reopen a contract closed to additional deliveries or close an open one. A "correction," simply enough, will allow adjustments to be made to contracts created in error. As you can imagine, I had to have a thorough understanding of how the firm prepared contracts to identify these modules.

Transactions created by BILL SALES and SETTLE PURCHASES may not be adjusted directly because they create source items (settlements and invoices) for the audit trail. However, the system must allow for indirect adjustments by providing a means to process debit and credit memos. In effect, a settlement or an invoice can be deleted if a debit or credit memo is issued for exactly the opposite amount of the original transaction.

The modules ISSUE PURCHASE ADVANCES and ISSUE ADVANCE BILLINGS also create transactions that can be adjusted. However, only the amount of the application of an advance billing to a final billing or an advance payment to a final settlement can be adjusted. The original entry that created the advance may not be adjusted because it too is a part of the firm's audit trail.

While I was examining the subordinate functions of the modules ISSUE CASH DISBURSEMENTS and PROCESS CASH RECEIPTS, I discovered that they'll have to do more than their names imply. Actually, what they need to do is keep the data stores PAYABLES and RECEIVABLES up-to-date. As a result, they must do more than process cash transactions. They must also allow adjustments to those data stores. Both functions must allow transfers of balances between data items for the same account and must permit discounts to be granted.

Now, all of the component functions of the modules of the preliminary structure chart modules have been identified. When you reach this point, it's again a mechanical process to expand the preliminary chart. Simply add a module for each newly identified function subordinate to the appropriate controlling module.

For instance, the module CREATE PURCHASE CONTRACTS in figure 6-2 will have four subordinate modules. Those are the four functions indicated in figure 6-3: create purchase contracts, fix prices on basis purchase contracts, correct purchase contracts, and change status on purchase contracts. Figure 6-4 shows how I expanded the chart to include these four functions, as well as all of the others indicated in figure 6-3.

You'll notice that I changed the names of each of the level-two modules in figure 6-2 when I expanded the preliminary chart. I did this to make the names of the

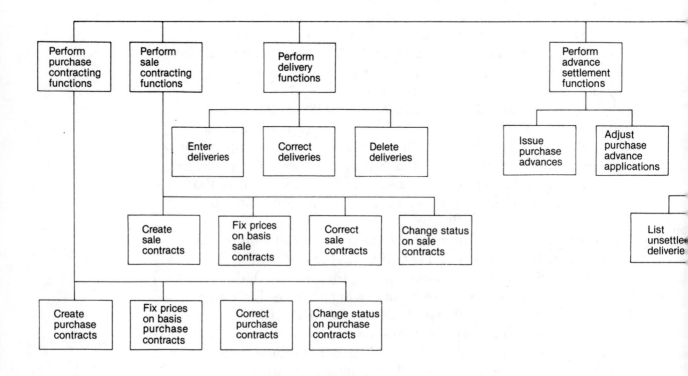

Figure 6-4   The expanded structure chart for the brokerage system's critical functions

modules as descriptive as possible. Since, for example, the module CREATE PURCHASE CONTRACTS in figure 6-2 was expanded to include subordinate modules that do more than just create contracts, the original name was misleading. So I changed it to PERFORM PURCHASE CONTRACTING FUNCTIONS, a more general but more accurate description of its function. Similar changes were required for the names of all of the level-two modules in the preliminary chart.

As you'll see in chapter 8, each of the low-level modules in figure 6-4 will probably become a separate application program when the system is completely designed. The final

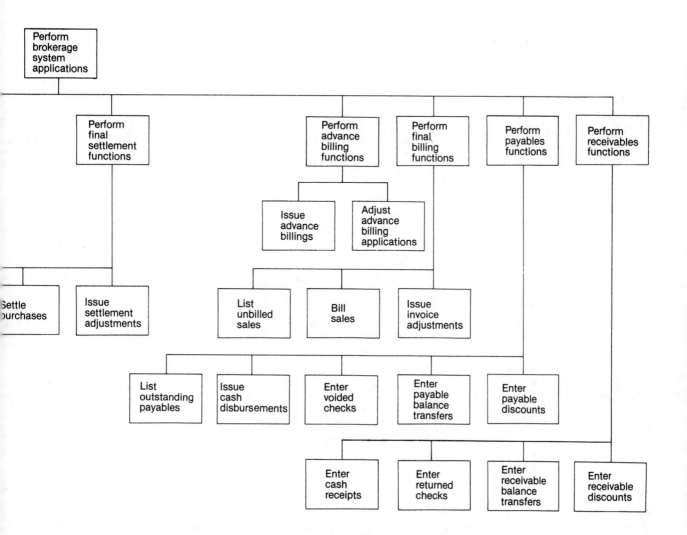

packaging of these structure chart modules into programs will come later. However, because independent functions have been identified and organized by the system structure chart, those packaging decisions will be easy. Often, in fact, the packaging of modules into programs is a trivial task. It's the underlying functions that are important.

Step 3: Add other required processes   Next, you need to review the items you chose *not* to include when you developed the model DFD for the system. For the brokerage

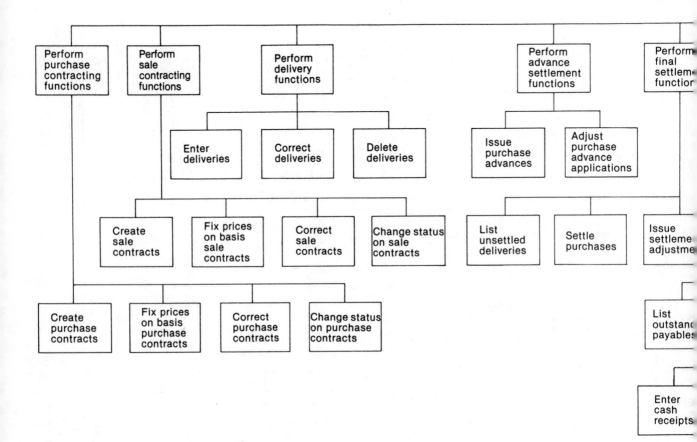

Figure 6-5 The structure chart for all of the brokerage system's required functions (alternative 1)

system, the functions I didn't include were (1) inquiries about transaction information and (2) extractions to prepare these outputs:

aged-receivables listing
aged-payables listing
mark-to-market report
position report
daily management report
monthly statements

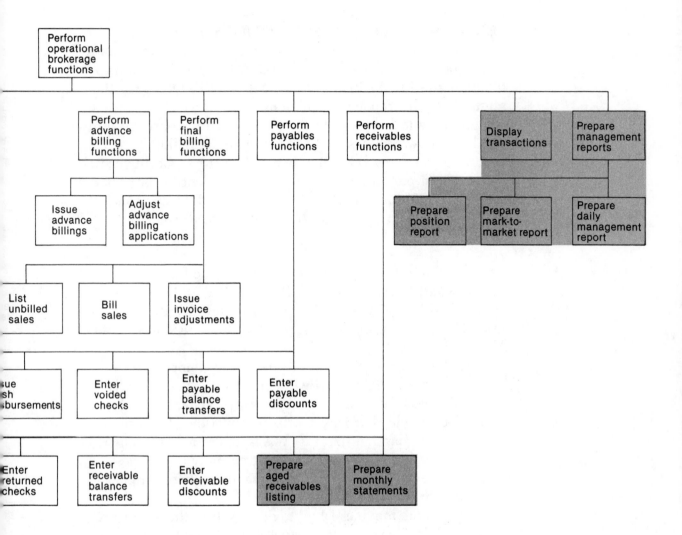

On review, I found that the aged-payables listing is really the same thing as the "outstanding payables" listing prepared in the module LIST OUTSTANDING PAYABLES. All of the other functions, however, needed to be added to the system structure chart.

To see how I integrated these additional functions into the system structure chart, look at the shaded portions of figure 6-5. Because the aged-receivables listing and the monthly statements are both drawn from the RECEIVABLES data store, it makes sense to include them subordinate to the

PERFORM RECEIVABLES FUNCTIONS module. To accommodate the remaining requirements, I added two second-level modules: DISPLAY TRANSACTIONS for the inquiry function and PREPARE MANAGEMENT REPORTS for the remaining three reporting functions.

As you add functions to a system structure chart, be sure you know how they would fit into the model DFD. Specifically, you need to know what data stores they'll access. For the most part, though, they're easy to identify. Be sure to keep the preliminary contents lists for those data stores current as you add functions to the system structure chart, however.

Step 4: Expand the other processes   Here, the three components of PREPARE MANAGEMENT REPORTS were what I started with, and I created their higher-level controlling module to organize them. As a result, I don't need to expand them any further.

On the other hand, I'm not sure at this point what DISPLAY TRANSACTIONS will consist of. When I finally specify the module (a process I'll describe in chapter 8), I'll get an accurate idea of its subordinate functions. Until then, I can defer its expansion. Because this is an extraction process, the subordinate modules that it will eventually be composed of (if it is indeed composed of subordinate modules) aren't operationally critical to the system.

Because the modules I had to deal with in this step were simple, I didn't have to spend much time trying to figure out what their components were. For other systems, though, you may find it difficult to expand the modules you added in step three. In other words, this step isn't always as easy as it happens to be for this brokerage system.

How to evaluate a system structure chart

As it stands, the structure chart in figure 6-5 is a useful document. It shows the functions that will have to be implemented in the new system. And it shows their relationships. However, it can be improved so it's easier to use and so it better reflects functional groupings.

At this point, each module at the end of each leg of the chart is a *work module*. That is, it performs a useful data processing function. In contrast, all other modules represent *control modules*. They determine when the lower-level modules will be executed. At this time, then, what you want to do is organize the structure chart so the work modules are controlled in the most logical and useful manner. How you do this depends on your own ideas about the system and, more important, on your users' needs, wants, and habits. As you reorganize the chart, the work modules won't change, but the control modules will.

When the control modules of a system are implemented, they take two general forms. First, they may be *menu programs*. A menu program lets the user decide which subordinate function to do next. Second, they may be *procedures*. A procedure consists of a chain of commands that causes one or more subordinate programs to be executed. In figure 6-5, for example, the prepare-management-reports module could be a procedure that causes all three subordinate functions to be performed. Or it could be a menu program that lets the user decide which of the three functions should be performed.

By the time you reach this point in system development, you'll probably have a thorough understanding of the system you're working on, but you should still work with the user. The user will appreciate your careful approach to planning the system, and you'll continue to get useful feedback on the design process that can save you costly troubles later.

Is the structure chart complete?   The most fundamental question you need to ask as you begin to evaluate a system structure chart is: Is it complete? Because you worked systematically from the model DFD, the chances are slim that you omitted any critical processes. But you need to check the items that you left off the model DFD to make sure all of them are reflected in the structure chart.

Are the modules grouped logically?   After you're sure that you haven't omitted anything from the system structure chart, you should look at the arrangement of what you have included. Ask yourself if all of the items on the chart are positioned logically. Because the chart was developed systematically

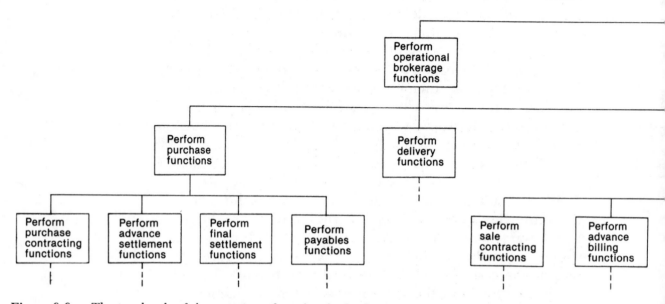

Figure 6-6   The top levels of the structure chart for the brokerage system (alternative 2)

from the top down, you probably won't find illogical groupings. But you will find processes that can reasonably be placed in more than one spot on the chart. In these cases, where you put a process may seem illogical to the user, and where the user would put the process may seem illogical to you. If that happens, you should defer to the experience and insight of the user.

Is the span of control of any module unreasonable?   The term *span of control* refers to how many modules a controlling module has subordinate to it. If a chart is arranged logically, span of control shouldn't be a problem. On the other hand, if you see modules with span of control problems, you can probably regroup their subordinates in a more logical way.

A span of control of one indicates you've tried to break a module down too far. Get rid of the single subordinate module, and let its superior perform its function. A span of control of two to five is probably acceptable. A span of con-

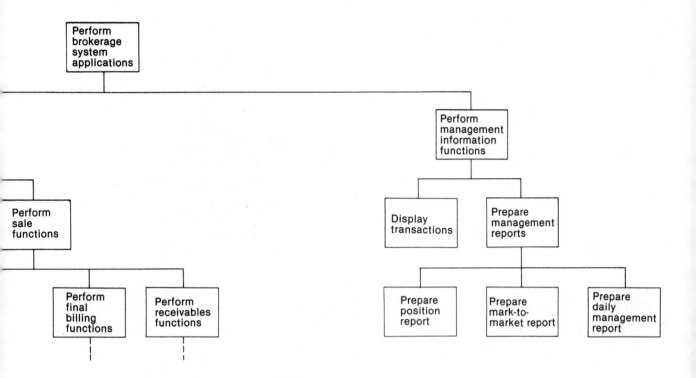

trol of six or seven may be acceptable. A span of control of
eight or more indicates you should seriously consider regrouping modules.

In figure 6-5, the top-level module must control 11 subordinate modules. Here, it would be reasonable to add extra
levels of control so no one level has such a broad span of control. Naturally, where you decide to put those extra levels and
how you group functions will vary. Let logic and the nature
of the business be the guiding factors.

Figure 6-6 shows how I regrouped the functions I identified for the brokerage firm. Since the firm's staff works with
the definite view that purchases and sales are separate, I
regrouped the processes to reflect that. As you can see,
regrouping required that I add several new mid-level
modules. In this figure, I left off some of the work modules so
you could concentrate on the control modules.

An alternative organization is in figure 6-7, and again I
left off some of the work modules. Here, I grouped similar
functions. Contracting functions are together, financial func-

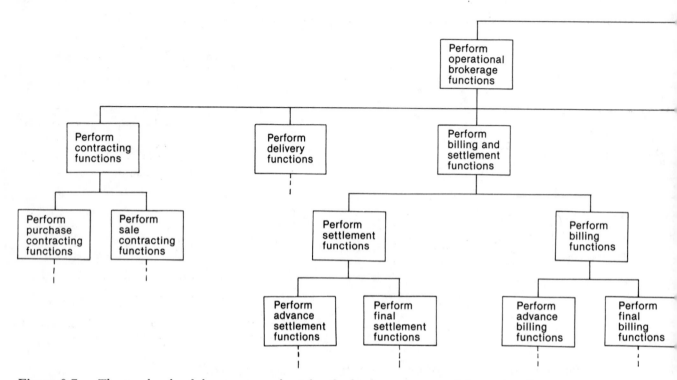

Figure 6-7   The top levels of the structure chart for the brokerage system (alternative 3)

tions are together, and so on. If you compare the charts in figures 6-5, 6-6, and 6-7, you can see that all of the work modules are the same, as they must be. The only differences in the charts is in how I chose to organize the work modules beneath controlling modules.

The importance I place on logical organization may seem overstressed, but it isn't. As you'll see, the structure chart will be the core of the new system's organization. I'll have more to say on this point later in the chapter. For now, I want you to understand that a system developed around a poorly organized system structure chart will be difficult and confusing for an operator to use. As a result, you must be careful to create a chart that's as logical and as sensible as possible. After you've organized the chart effectively, the next step is to redraw it and label its parts in a more useful way.

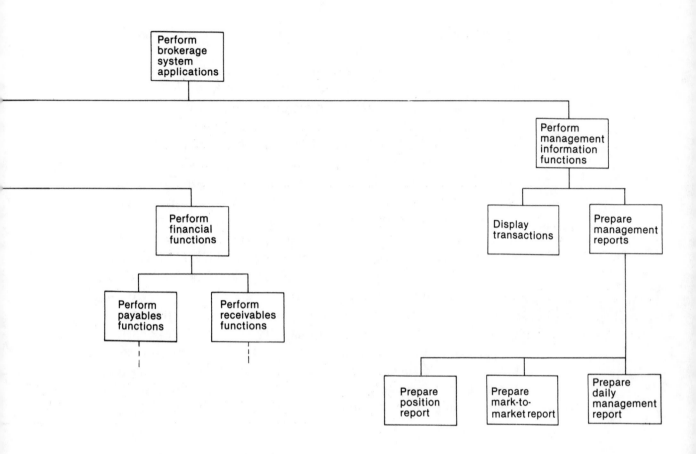

How to draw a system structure chart

Now I'd like to give you some suggestions for drawing a
system structure chart that will produce a document that's as
useful as possible. As you can see in figure 6-5, a typical
system structure chart will have many modules. To keep a
system structure chart from being too confusing, you should
adopt a standard method of formatting it and identifying its
modules.

How to draw multiple-page charts   You should realize by
now that you can't fit the entire structure chart for even a
modest system on a single piece of paper. In fact, the chart in
figure 6-5 represents only a subset of the entire brokerage

system. For most systems, then, you'll need to use several pages for your system structure chart.

Figure 6-8 is a revision of the structure chart in figure 6-6, redrawn on several pages. What I try to do is fit as much as possible on each page without crowding. I also try to keep all processes shown on one page at about the same level in the hierarchy so the chart isn't misleading. Beneath each module that's expanded elsewhere, I indicate to what page the reader should turn to find the continuation. On the continuation, I refer back to the page on which the module first appeared.

This process continues down as many levels as necessary to accommodate all of the modules on the chart. In this example, I used 11 pages. Presenting the system structure chart on several pages doesn't allow you to look at the whole logical structure in one view as you can with the chart in figure 6-5, so that's an advantage you lose when you go to a multiple-page chart. Although you can use large sheets of paper for your charts, anything bigger than 11 by 17 inches (the size of two letter-size sheets side by side) is awkward. Sheets larger than 11 by 17 are hard to manage when you're drawing on them, won't fit most copiers, and are unwieldly to file.

How to name the modules   As you've already seen, I changed some module names as I expanded the system structure chart. I tried to give each module a name that accurately describes what it does. For a control module, this means trying to summarize the functions of all of its subordinate modules in a brief name. By necessity, the closer to the top of the chart a module is, the more general its name must be. As a guideline, try to format each name using a verb, an adjective or two, and a noun. An example is PREPARE MANAGEMENT REPORTS.

How to number the modules   No matter how well a name describes a module, referring to the module by name can be cumbersome. So I suggest that you give each module a number as well as a name. The numbering scheme I use consists of a code that identifies a functional portion of a system followed by a numeric identifier. For instance, the number BCK1000 might refer to module 1000 in a portion of a system that does backup (BCK).

In part 1 of figure 6-8, I numbered the top-level module SYS0000. Beneath it, I numbered the top-level brokerage module BF0000 and the top-level management information module MGT0000. Subordinate to both of these, I retained the same alphabetic prefix as I numbered the modules. My numbering scheme reflects the hierarchical structure of the modules. For instance, PERFORM PURCHASE CONTRACTING FUNCTIONS is BF1100, subordinate to BF1000, subordinate to BF0000.

Later in the development process, you'll find that you can use the module numbers as program names or procedure names. If you do, the system structure chart will direct you to all of the programs and procedures of the system. Just be sure that the numbering scheme you use is compatible with the naming requirements of your system. As a general rule, limit yourself to eight-character names that begin with a letter.

An advantage of numbering modules in this way is that it's easy to add individual modules or an entire subsystem to the chart. To add a new module, first position it on the chart where it logically belongs. Then, give it a number that falls between those of the modules on either side of it. To illustrate, suppose I realize during the development of the brokerage system that I need to add a new function to reprint copies of purchase contracts. If you look at the system structure chart in part 2 of figure 6-8, you'll see that all of the modules related to purchase contracts are subordinate to module BF1100. So that's the logical place to add this function. Then, if I decide to insert it between CREATE PURCHASE CONTRACTS and FIX PRICES ON BASIS PURCHASE CONTRACTS, I assign it a number between BF1110 and BF1120 (probably BF1115). If I choose to add it after CHANGE STATUS ON PURCHASE CONTRACTS, I assign it a number like BF1150.

To add a new subsystem is just as easy. Suppose, for example, that I want to add a general ledger subsystem to the system structure chart for the brokerage system. First, I add a new second-level module at the same level as BF0000 and MGT0000 in part 1 of figure 6-8. Then, I assign it a number with a new prefix that indicates its broad function followed by 0000, as in GL0000. Finally, subordinate to that new module, I add all the new general ledger modules. In the next chapter, I'll present another example of expanding the system structure chart to include a new subsystem.

Brokerage system structure chart
page 1 of 11

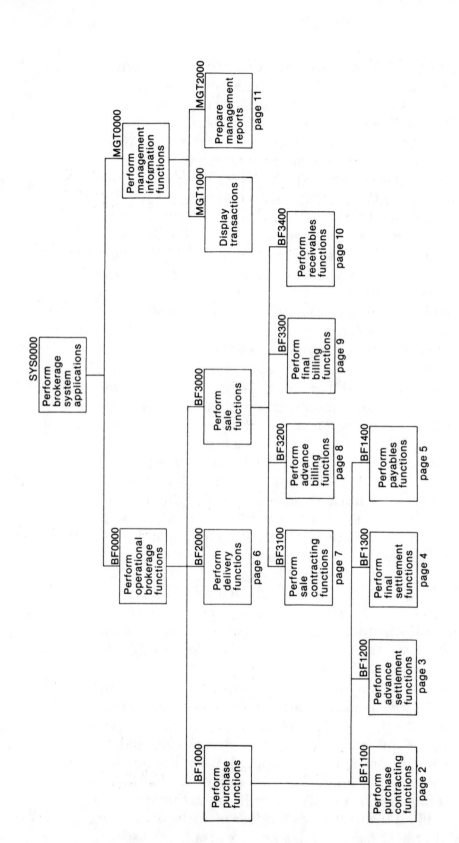

Figure 6-8 The refined system structure chart for the brokerage system (part 1 of 11)

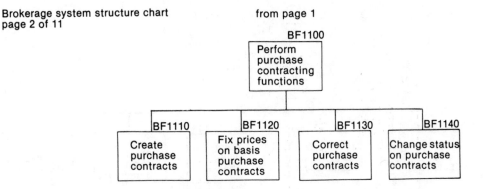

Figure 6-8   The refined system structure chart for the brokerage system (part 2 of 11)

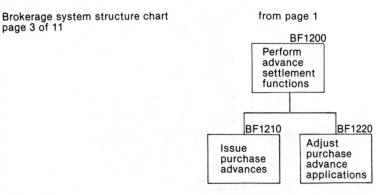

Figure 6-8   The refined system structure chart for the brokerage system (part 3 of 11)

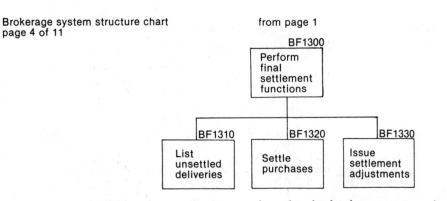

Figure 6-8   The refined system structure chart for the brokerage system (part 4 of 11)

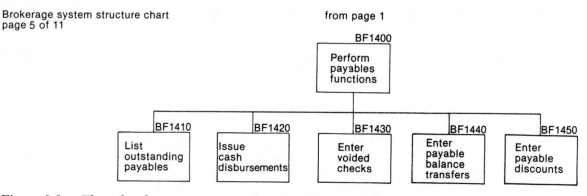

Figure 6-8   The refined system structure chart for the brokerage system (part 5 of 11)

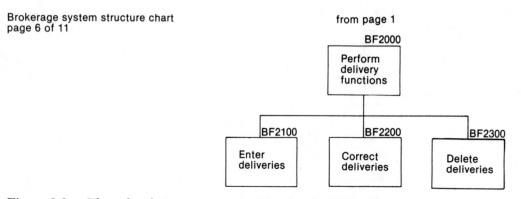

Figure 6-8   The refined system structure chart for the brokerage system (part 6 of 11)

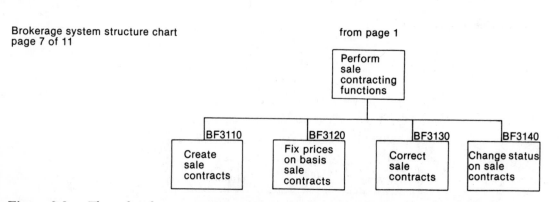

Figure 6-8   The refined system structure chart for the brokerage system (part 7 of 11)

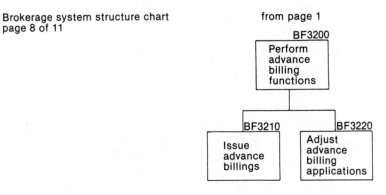

Figure 6-8   The refined system structure chart for the brokerage system (part 8 of 11)

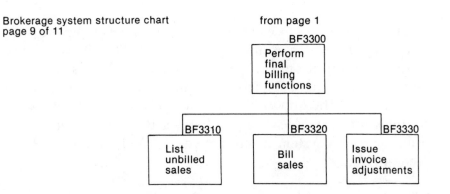

Figure 6-8   The refined system structure chart for the brokerage system (part 9 of 11)

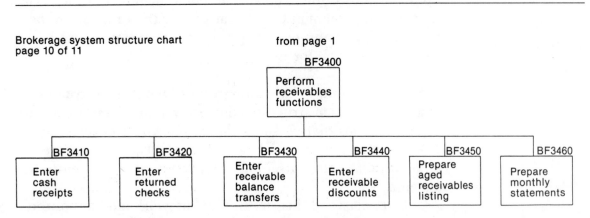

Figure 6-8   The refined system structure chart for the brokerage system (part 10 of 11)

Brokerage system structure chart
page 11 of 11

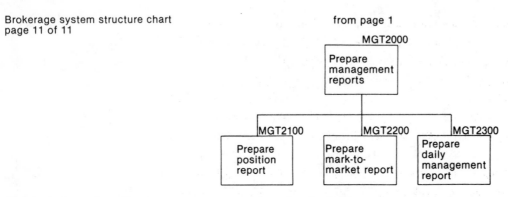

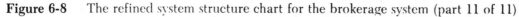

Figure 6-8   The refined system structure chart for the brokerage system (part 11 of 11)

How to implement a menu-driven system

Chances are, if you're working on a system development project today, it will be an interactive system. That leaves plenty of room for variation, though. At one end of the scale, you might be planning a single-user application for a personal computer. At the opposite end, you might be developing a complex, nationwide system. Whatever the scope of your project, if your system is interactive, it's likely to be menu-driven.

What is a menu-driven system?   Basically, a *menu-driven system* is one that presents the user with lists of processing options. From each list, or menu, the user can select one option for execution. The alternative to menus is to force the user to enter command lines that specify the functions to be performed. A menu-driven system is less error-prone and is easier for an operator to use.

How to develop a menu structure from the system structure chart   The system structure chart shows the hierarchy of program functions that make up the system under development. Accordingly, you can work directly from it to create a menu structure that can control and direct users as they operate the system.

To understand this, recall that the work modules of the system structure chart represent application programs. In con-

trast, the control modules are either menu programs or procedures. In a typical interactive system, most of the control
modules are implemented as menu programs.

Each menu program is executed through the appropriate
selection from the menu superior to it and allows selection of
any of the modules subordinate to it. Those subordinate
modules may be application programs or procedures or lower-
level menus.

To illustrate the operation of a menu-driven system, look
back to the system structure chart for the brokerage system in
figure 6-8. Assume that the top-level module, SYS0000, is implemented as a menu program called SYS0000 that allows the
user to perform the two functions subordinate to it. More
specifically, the program SYS0000 displays a "menu" of selections, permits the user to choose one of the displayed selections, and causes the selected function to be executed. Part 1
of figure 6-9 shows the sort of menu the program SYS0000
might display and shows the related modules of the system
structure chart. Both of the subordinate functions can be
selected from the menu.

In addition to these two options, the menu in part 1 of
figure 6-9 offers a third that allows the user to log off the
system. Each menu you include in your system should have
an "escape" like this, a selection to return to the menu that
called it. Because the top-level module of a system (often
called the *master menu*) has no calling menu, the only escape
from it is to log off the system.

Each of the selections displayed on the master menu can,
in turn, display another menu. If you were to select option 1
from the menu in part 1 of figure 6-9, you'd enter the
brokerage-functions subsystem. Logically, you'd move down
in the system structure chart from module SYS0000 into
BF0000. Physically, the program SYS0000 would cause the
program BF0000 to be loaded and executed. The program
BF0000 would display another menu and permit the user to
execute any of the functions subordinate to the module
BF0000 on the system structure chart. Part 2 of figure 6-9
illustrates what that menu might look like and shows its relationship to the system structure chart. Here again, I included
a selection to escape from the displayed menu and return to
the calling menu, SYS0000.

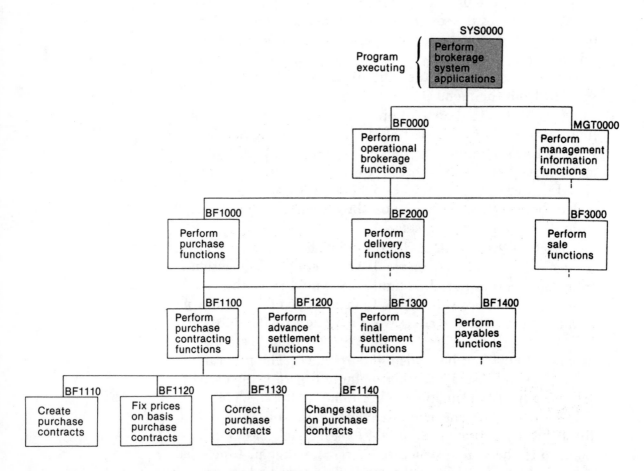

Figure 6-9   The menu structure for the brokerage system: the menu for SYS0000 and the related portion of the system structure chart (part 1 of 4)

If the user selects number 1 from the previous menu, the program BF0000
will be executed and will display this screen:

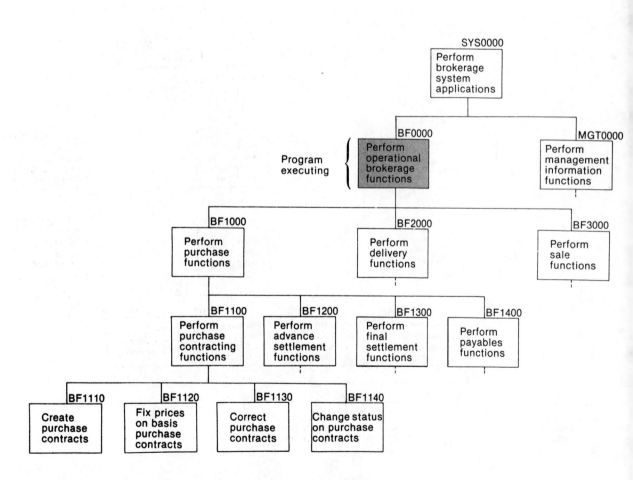

Program to be
executed with a
specific selection:

```
OPERATIONAL BROKERAGE FUNCTIONS

1. PERFORM PURCHASE FUNCTIONS
2. PERFORM DELIVERY FUNCTIONS
3. PERFORM SALE FUNCTIONS

ENTER YOUR SELECTION: __

PRESS CLEAR TO RETURN TO  PREVIOUS MENU
```

BF1000
BF2000
BF3000

SYS0000

SYS0000

Perform
brokerage
system
applications

BF0000

Program
executing

Perform
operational
brokerage
functions

MGT0000

Perform
management
information
functions

BF1000

Perform
purchase
functions

BF2000

Perform
delivery
functions

BF3000

Perform
sale
functions

BF1100

Perform
purchase
contracting
functions

BF1200

Perform
advance
settlement
functions

BF1300

Perform
final
settlement
functions

BF1400

Perform
payables
functions

BF1110

Create
purchase
contracts

BF1120

Fix prices
on basis
purchase
contracts

BF1130

Correct
purchase
contracts

BF1140

Change status
on purchase
contracts

Figure 6-9   The menu structure for the brokerage system: the menu for BF0000 and the related
portion of the system structure chart (part 2 of 4)

If the user selects number 1 from the previous menu, the program BF1000
will be executed and will display this screen:

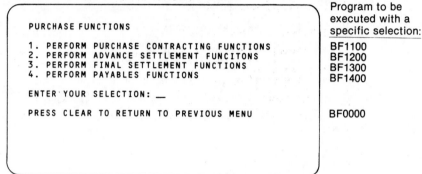

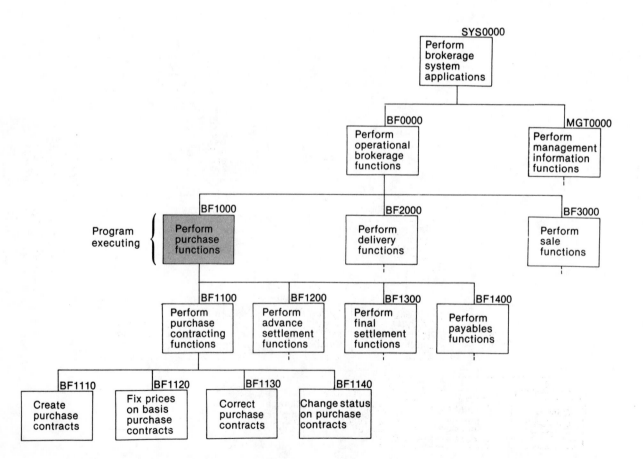

Figure 6-9   The menu structure for the brokerage system: the menu for BF1000 and the related
portion of the system structure chart (part 3 of 4)

If the user selects number 1 from the previous menu, the program BF1100
will be executed and will display this screen:

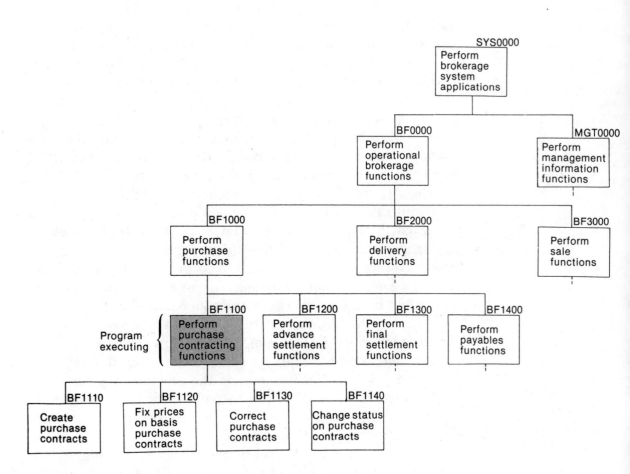

```
PURCHASE CONTRACTING FUNCTIONS

1. CREATE PURCHASE CONTRACTS
2. FIX PRICES ON BASIS PURCHASE CONTRACTS
3. CORRECT PURCHASE CONTRACTS
4. CHANGE STATUS ON PURCHASE CONTRACTS

ENTER YOUR SELECTION: _

PRESS CLEAR TO RETURN TO PREVIOUS MENU
```

Program to be
executed with a
specific selection:

BF1110
BF1120
BF1130
BF1140

BF1000

Figure 6-9   The menu structure for the brokerage system: the menu for BF1100 and the related
portion of the system structure chart (part 4 of 4)

You can continue to implement each box of the system structure chart as a menu down to the point where the chart's modules represent processing programs. Parts 3 and 4 of figure 6-9 illustrate how a user can move deeper and deeper into the menu structure until an application program is encountered at the lowest level of the structure. For example, by selecting option 1 from the menu in part 4 of figure 6-9, the user would enter an application program to create purchase contracts.

Why logical organization and span of control are critical to a menu-driven system   Earlier in this chapter, I stressed the importance of developing a system structure chart that's logically organized and that doesn't have modules with unreasonable control spans. Now that you understand how easily a menu structure can be derived from the system structure chart and how the user depends on menus to move through a system, I think you can recognize why organization and span of control are so important.

Figure 6-10 presents two master menus that have some serious span of control and organizational problems. Both are from application packages for minicomputer systems. The top menu is from an accounts payable package, and the bottom menu is from a medical billing and accounts receivable package. If you analyze these menus, you should see some problems that make them difficult to use.

First, both of these menus have 14 options from which the user must select one. I think you'll agree that 14 is too many. Not only is the operator forced to find the proper function in a long list, but the chance for an entry error increases as the number of processing options increases.

Second, the span of control problem is aggravated by the poor organization of these menus. Functions that seem to belong together should be grouped together, but in these menus they aren't. For instance, in the accounts payable menu, selections related to the system's vendors are numbers 1, 4, and 9. So wouldn't the menu be easier to read and use if those options appeared in a group?

Third, the level of the selections ranges from a high-level menu to a work module. In the medical billing and accounts receivable menu, the first selection, DAILY PATIENT

```
ACCOUNTS PAYABLE


    PLEASE SELECT

(1)  VENDOR FILE MAINTENANCE
(2)  NEW A/P TRANSACTION PROCESSING
(3)  PRINT A/P OPEN ITEM REPORT
(4)  VENDOR ACCOUNT INQUIRY
(5)  A/P OPEN ITEM ADJUSTMENT
(6)  PRINT CASH REQUIREMENTS REPORT
(7)  PAYMENT PREPARATION AND CHECK PRINTING
(8)  PRINT A/P DISTRIBUTION TO G/L REPORT
(9)  PRINT VENDOR ANALYSIS REPORT
(10) CHECK RECONCILIATION
(11) JOB FILE MAINTENANCE
(12) PRINT JOB DISTRIBUTION REPORT

(15) SELECT ANOTHER COMPANY TO PROCESS
(16) END OFF ON ACCOUNTS PAYABLE PROCESSING
```

```
MEDICAL BILLING AND A/R

                  DAILY INPUT MENU          06/03/83

      PLEASE SELECT THE OPTION YOU WANT-

 1:  DAILY PATIENT TRANSACTIONS
 2:  RECEIPTS/ADJUSTMENTS
 3:  PATIENT MASTER UPDATES
 4:  INQUIRY-ACCOUNT STATUS
 5:  INQUIRY-ACCOUNT HISTORY
 6:  PATIENT RECALL NOTICE UPDATES
 7:  RECALL MESSAGE UPDATES
 8:  INSURANCE COMPANY MASTER UPDATES
 9:  PATIENT BALANCE FORWARDS (INITIAL SET-UP ONLY)
10:  BILL-FUTURE BILLING (FROM OPEN ITEM FILE)
11:  RUN SPECIAL PROGRAM/OTHER INPUT REQUEST
12:  PROCESSING/MISC MASTER UPDATE MENU
13:  SIGN OFF SYSTEM
14:  CONTRACTS/PRE-AUTHORIZED
```

Figure 6-10   Two master menus with confusing organizations

TRANSACTIONS, leads to other menus that eventually take
the user to several critical entry programs. DAILY PATIENT
TRANSACTIONS, therefore, is relatively high in the system's
structure (if there is one). In contrast, the sixth selection,

PATIENT RECALL NOTICE UPDATES, is a function that's an application program. Nevertheless, it's located in the menu structure at the same level as DAILY PATIENT TRANSACTIONS. The result is that the menu is confusing. If you develop a menu structure from a system structure chart that's logical, the menu structure will be logical, and you'll encounter few organizational problems like this.

Finally, these menus also illustrate how important it is to assign modules names that are clear and unambiguous. To illustrate, compare the descriptions of selections 6 and 7 in the medical menu. Can you tell what the difference between PATIENT RECALL NOTICE UPDATES and RECALL MESSAGE UPDATES is? I can't. Selection 14, CONTRACTS/PRE-AUTHORIZED, is another inadequate name.

In contrast to the menus in figure 6-10, well designed menus make learning how to operate a system an easier task for users. In fact, a logical menu structure helps make a system self-instructional. This in turn reduces the amount of time you have to spend developing user manuals. So both your productivity and your user's productivity are increased.

How to use a menu structure to limit access to a system   As shown in figure 6-8, the menu structure I've set up for the brokerage system gives each user unrestricted access to all programs of the system. For most systems, though, that's unacceptable. Fortunately, however, it's easy to make modifications to the menu structure that allow limited access to the system for specific users or classes of users.

The basic idea here is that users shouldn't see processing options they aren't authorized to do. In the brokerage system, for example, there are four classes of users with respect to what functions they may perform: managers, new clerks, experienced clerks, and brokers. Since the new clerks are most limited, they'll never be offered selections they're unauthorized to perform. In contrast, managers will be offered all of the selections of the system because they have unlimited access to it.

The managers' menu structure, then, is reflected in part 1 of figure 6-11. This is the top portion of the system structure chart for the brokerage system, to which managers have unrestricted access.

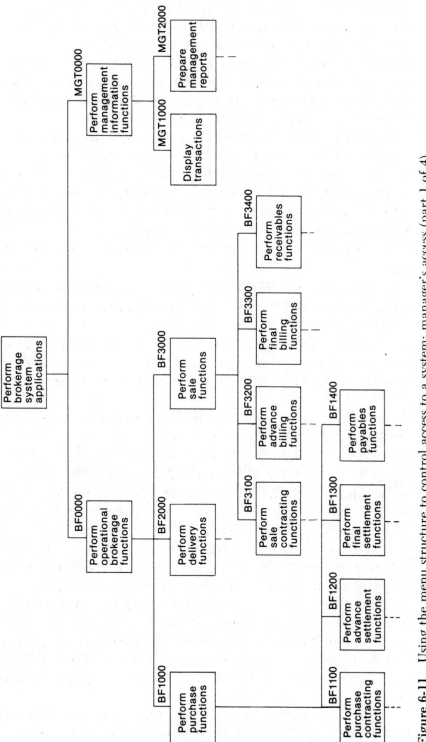

Figure 6-11 Using the menu structure to control access to a system: manager's access (part 1 of 4)

On the other hand, the managers don't want all users to have access to all functions, particularly those that involve fund transfers (payables and receivables) and those that summarize the firm's performance (management reports). As a result, when a new clerical employee is hired, he is given restricted access to the system. A new user is allowed to do only contracting and delivery functions and may have access only to inquiry functions in the management information subsystem.

If you look at part 2 of figure 6-11, you can see that I've shaded those parts of the system structure chart to which a new user may have access. To provide this security, I set up a separate menu system for new users (automatically entered when a user signs on to the system) that allows only the selections within the shaded section of the system structure chart. Although this may sound like a major undertaking, it really isn't. Menu programs are typically simple, so they're easy to create and modify.

As new clerks gain experience and the confidence of management, they're allowed to prepare final settlements and invoices. The shaded area in part 3 of figure 6-11 encloses the system structure chart functions that make up this subset of the complete menu system. Still, the experienced clerk doesn't have access to financial and management reporting functions.

The fourth class of employee is the broker. Brokers have no need to access any of the operational brokerage functions. However, they do need to be able to inquire into transactions and to prepare management reports. Accordingly, the menu structure for the firm's brokers consists of those functions within the shaded area in part 4 of figure 6-11.

You can see in figure 6-11 that when a subset of the system is implemented for a class of users, some of the levels in the chart may become superfluous. For example, in part 2 of figure 6-11, the modules BF1000, BF3000, and MGT0000 are unnecessary. MGT0000 is also unnecessary in part 3. When this is the case, you should omit these modules from the menus. As you can imagine, it's foolish to force a user to pass through a menu that has only one selection.

How to use a menu structure to control time-dependent processing   In many applications, processing is time-dependent.

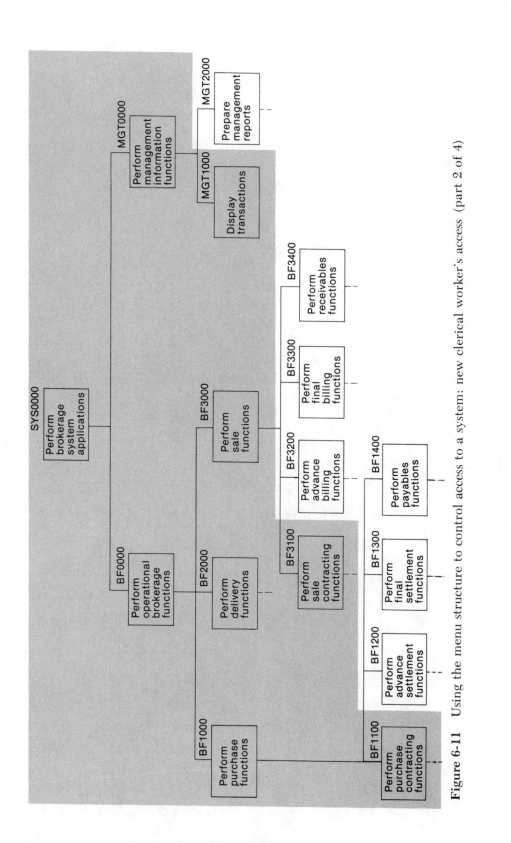

Figure 6-11 Using the menu structure to control access to a system: new clerical worker's access (part 2 of 4)

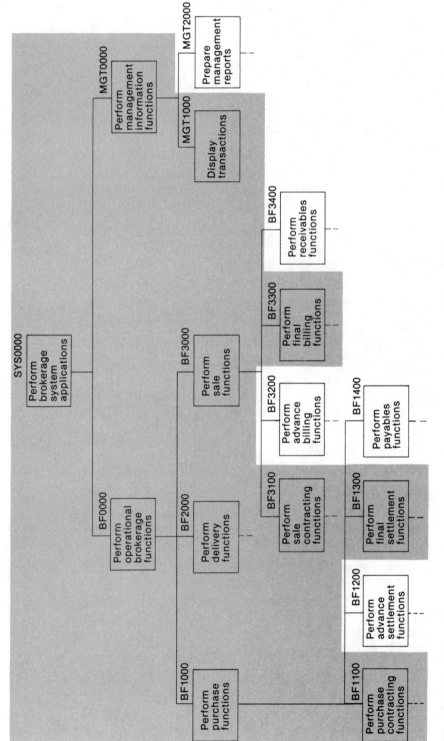

Figure 6-11   Using the menu structure to control access to a system: experienced clerical worker's access (part 3 of 4)

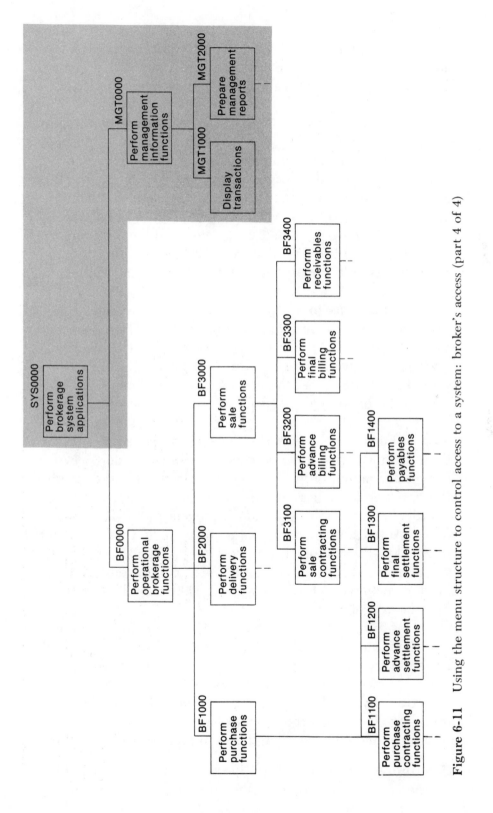

Figure 6-11 Using the menu structure to control access to a system: broker's access (part 4 of 4)

In the brokerage system, for example, it's necessary to prepare a position and a mark-to-market report at the end of every month, *before* transactions for the new month are processed. This timing, of course, is critical to the accuracy of the system.

In many systems, though, the operator is responsible for requesting period-end reports. So what happens if he forgets to request the reports and proceeds with the next period's work? If the period-end reports are prepared later, they'll be inaccurate. Then, to prepare accurate reports, the database has to be restored to the state it was in before the new-period transactions were processed. That means either restoring the affected part of the database from backup copies of it or reversing all of the new-period transactions using application programs.

The best way to avoid these irritating corrections is not to allow mistimed processing in the first place. The easiest way to do this is to write menu programs that don't allow an operator to enter transaction-processing modules for a new period until period-end processing has been done.

In the brokerage system, then, the master menu (SYS0000) shouldn't allow the user to execute the second-level menu BF0000 in a new period until the required reports for the previous period have been prepared. The easiest way for the master menu to do this is to access control data that gives the date when the last period-end reports were prepared. It can then compare this date to the current date. If the two dates are in different periods, the menu should restrict access to BF0000. Of course, the application module that prepares the period-end reports must update the control data so the master menu can allow access to BF0000 once the reports have been prepared.

This is only a simple example of how a control module can eliminate errors related to time-dependent processing. But I'm sure you can think of others. Time-dependent processing often occurs on a daily basis and often involves more than just reporting. By checking control data that indicates when functions have been performed, your control modules can make sure that functions are done in the proper sequence.

How to use a menu structure to control backup   Backup is another requirement that can be integrated into a menu struc-

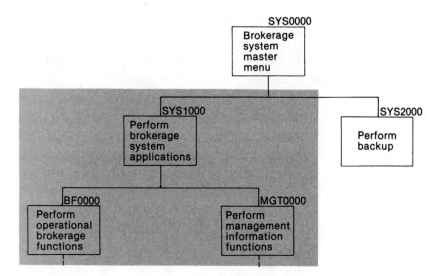

Figure 6-12 The structure chart for the brokerage system modified to include a backup function

ture. This is more of a consideration for small systems than it is for larger mainframe systems. In a large- or medium-sized shop, backup is the responsibility of the systems staff, and the end user typically has nothing to do with it. On a small system, however, backup is usually the responsibility of the end user.

When backup is the responsibility of the user, it's critical that it be done regularly. As a result, it's a special case of time-dependent processing. And your system menu structure can control the timing of backup just as it can control other time-dependent processing. The only thing special about controlling backup from the menu structure is that access to the entire system, not just a part of it, is restricted when backups haven't been done. That means that you need to add a level to the system structure chart to accommodate the backup function.

Figure 6-12 shows the top levels of a system structure chart for the brokerage system that includes a backup function. The top levels of the original structure chart in figure 6-8 is within the shaded area. You'll notice that I changed the number of the original top-level module from SYS0000 to SYS1000 to reflect the fact that I added another level to the chart. In this system, module SYS0000 can check to make sure that backup module (SYS2000) has been executed for the day before allowing anyone access to module SYS1000.

How to use the system structure chart

After you create a structure chart for a system, it becomes a practical development document. Because users can easily understand it, it provides a basis for discussion. In addition, it can help guide the development process in several ways.

How to get the user's ideas about the structure chart   As I stress throughout this book, one of the most important tasks of the system designer is to communicate effectively with the user during all phases of the system development process. The user must understand what you're doing if he's to provide useful information. Conversely, you must fully understand the requirements and wishes of the user. At this point, then, you should discuss the system structure chart with the user.

When you discuss the chart with a user, you should stress that it represents functions in a hierarchical arrangement, not details of processing. Further, you should point out that each leg, or subsystem, is relatively independent of the others. As a result, you can discuss each leg of the chart as an independent unit. As you do, prompt the user to tell you if you've left anything off the chart. Also, make sure that the functions you have provided are really necessary. Get the user to think of what he needs that isn't provided by the functions shown on the chart.

When the user does make suggestions or asks for additions, deletions, or modifications, you should mark your copy of the chart during the discussion. Then, when you're back at your desk, you can create a new chart that includes the changes made by the user. When you're done, you can discuss the new version of the chart with the user. If you continue this process until the user has no more suggestions for changes, you can feel sure that you've provided most of the functions that the system requires. Usually, this is a brief process because most requirements were identified when the system DFDs were developed.

How to use a chart to plan a development sequence   In many instances, the users of a new system will want to have certain parts of the system operating as soon as possible, while other parts won't be so important to them. For the brokerage system, the users were anxious to improve their billing procedures. As a result, it made sense to develop the PERFORM FINAL BILLING FUNCTIONS module and its subordinates first. Then, part of the system could be up and running, offering its benefits to the users, as rapidly as possible. Because the system structure chart emphasizes independent functions and is modular, you can easily identify the programs that need to be developed first to meet a user's needs.

Similarly, you can use a system structure chart to select a usable subset for initial implementation. Since a system structure chart will usually include modules that aren't critical to a system's functions, you can defer development of these modules while you concentrate on the critical modules. By doing so, you can implement a working subset of the final system more rapidly than if you developed the complete system. Then, as the less critical modules are developed, they can be added to the working system.

How to use a chart to monitor development progress   A master copy of the final system structure chart can graphically show how the development of a system is progressing. If you mark out or color completed modules as they're finished, you can tell in an instant how close any portion of the system is to completion. You can also identify problem areas in the development process. Perhaps most important, posting a copy of the system structure chart in the programmers' work area and marking off completed modules of the system can give the staff a feeling of accomplishment and progress in a process that all too often seems endless.

Discussion

Creating the system structure chart is a critical part of the system development process. As I've just explained, the system structure chart and the menu structure based upon it provide the framework for developing the final system. In chapter 8, I'll show you how to specify what each application module of the system structure chart will do.

Although the examples in this book show you how to develop an interactive system, you should realize that you can use a structure chart for designing batch systems too. Within an interactive system, of course, there are often subsystems that run on a batch basis. In this case, the control module of the subsystem is a procedure that calls other procedures or programs until all of the batch programs in the subsystem have been executed. If you try using structure charts to design batch systems, I think you'll find that they're far more effective than traditional design documents like system flowcharts.

To give you more perspective about this phase of development, I want you to know that I rate the system structure chart as the most important development document recommended in this book. As soon as we started using structure charts, both our productivity and our users' satisfaction improved. And we started using structure charts a few years before we started using DFDs. As you've seen, structure charts are particularly useful when designing interactive systems. But I recommend the structure chart as the primary development document for any kind of computer system. By using one, you'll make sure that you've provided for all required functions and that you've organized these functions within a logical control structure.

Terminology

system structure chart
process
function
module
work module
control module
menu program
procedure
span of control
menu-driven system
menu
master menu

Objectives

1.   Given a model DFD and information about other required functions, create an effective system structure chart.

2.   Explain how to develop a menu structure from a system structure chart.

3.   Explain how to use a menu structure to control access to a system, to insure that time-dependent processing is done at the right time, and to provide for backup.

4.   Explain how to use a system structure chart to get the users' ideas, to plan a development sequence, and to monitor development progress.

Chapter 7

How to design a system's database

In this chapter, I'll show you how to design a system's database. Although I've divided system design into process specification and database design in this book, you should realize that this division is artificial. At some point during design, you've got to look at how process requirements affect data requirements, and vice versa. In fact, as I explained in chapter 1, the two development phases normally overlap. In other words, many of the decisions you make regarding database design are based on knowledge you only get by examining how particular processes will function. As a result, process specification, as described in the next chapter, is normally done concurrently with database design.

This chapter has to be somewhat general, because the more detailed design becomes, the more specific it is to a system configuration and user application. In particular, the wide variety of ways you can implement a database today keeps me from setting down a foolproof method for designing an effective database. To design a database, you'll need specific technical knowledge about the capabilities of the system on which you'll be working. What I *can* offer is a general procedure that will help you through the creative process of database design.

---

1. Identify omitted data stores.

2. Draw a data access diagram to show required access paths.

3. Draw a data hierarchy diagram to clarify data relationships.

4. Evaluate and refine the database plan.

5. Decide how to package the system's data elements.

6. Document the physical database design.

Figure 7-1   The six steps for designing a system's database

Figure 7-1 shows my recommended procedure for designing a database. As I warned you, these steps are general. But if you apply your knowledge of your computer within this procedure, I'm confident that you'll approach database design more systematically than you ever have before. In a minute, I'll explain and illustrate each of these steps. But first, I'd like to make a few comments about knowing the capabilities of your computer system.

Knowing the capabilities of your computer system

When you develop a system for a specific computer, the capabilities of that computer affect your design, in particular, your database design. As a result, you must know the computer's capabilities, both in terms of hardware and software.

Hardware capabilities that affect design are factors such as disk storage capacity, disk speeds (both access and transfer), and internal speeds. These factors can affect the response times at the workstations of an interactive system, the run times for batch jobs, the size limitations for files, and so on. If the hardware doesn't have the capacity to run the jobs of the system at satisfactory speeds and with satisfactory response times, you need to recognize the problem as early in the analysis process as possible and develop solutions for it. This kind of analysis, however, is beyond the scope of this book.

The software capabilities of a system have a more direct effect on the design of a database. If you work on a system where a database management system (DBMS) is installed,

you need to know its specific characteristics. Is it a hierarchical, network, or relational system? What limitations are
imposed on the relationships your DBMS can maintain? What
are the performance implications of specific implementations?
And so on. For example, simulating network data structures
under a hierarchical DBMS can impose significant processing
overhead and introduce operational inefficiencies. So you
must know what the alternatives available to you are if you
want to design an efficient database.

If you don't have database software, you must be
thoroughly familiar with the file-handling capabilities of your
system. Almost all systems today provide for indexed file
handling. But do the indexed capabilities provide for alternate
keys? And if so, do they provide for non-unique alternate
keys? Factors like these can have a significant effect on the
design of your database.

Here again, the database software and file-handling
capabilities of specific systems are beyond the scope of this
book. But you must realize that you can't design an effective
physical database without knowing these capabilities. With
that as background, let me present the six steps that I recommend in my procedure for designing a database.

Step 1: Identify omitted data stores

You already have preliminary contents lists for the model
DFD's data stores. As you know, those data stores will be the
critical data stores of the system. That's not to say, however,
that they're the *only* data stores of the system. Because
reference and archive data stores were omitted from the
model DFD, you have to add them now if you want the
database to be complete and accurate.

Identifying reference data stores   A *reference data store* is a
collection of data that's used by processes but is not updated
by them. In a sense, then, a reference data store is not required for the operation of a system. However, the advantages one offers can be significant. In general, a reference
data store serves two primary functions: (1) to reduce data
redundancy and (2) to provide standardized codes.

When you reduce data redundancy in a system, you make more efficient use of storage. At the same time, you insure the integrity of the database because only one occurrence of a structure needs to be updated when changes are necessary. Then, you eliminate the possibility that one occurrence of a structure like customer address has been changed but another occurrence hasn't been changed.

At this time, then, you should examine the contents lists you've already prepared. When you find data structures that occur in several lists, you should consider implementing these structures as separate, reference data stores.

In the brokerage system, for example, the names and addresses of buyers and sellers occur throughout the database. As a result, I decided to create separate data stores called BUYERS and SELLERS that contain those names and addresses. Then an occurrence of the name and address structure for a buyer is identified by a code that's the access path to the data store BUYERS. That code will appear in all related data stores in place of the complete name and address.

When you decide to use a reference data store, you need to create a contents list for it. Because the contents lists are used in the last steps of database design, you need to keep them complete and up-to-date. As a result, you also need to update the contents lists of all other data stores that contain elements that will be included in the new reference data store. In particular, you must replace those elements with the identifying code for the new reference data store. Figure 7-2 illustrates this type of change using the contents lists for the data stores SALE CONTRACTS and BUYERS. In this case, I also need to replace the occurrences of buyer name and address in all other data stores with the buyer code.

When you encounter a data element that's widely used in a system, you should consider standardizing it throughout the system. If you decide to do so, you store the standardized codes in a reference data store. Typically, you should standardize the values of a data element if any system functions require summaries based on that data element.

A data element that appears throughout the brokerage system is commodity. As a result, I decided to standardize the commodity identifier in all data structures. I made this decision because several reports required all data structures for a

Original contents list for the data store SALE CONTRACT:

Data structure: Sale contracts

Group	Repetitions	Components	Comments
Y		Sale contract number Buyer name and address Commodity Sale contract price Open/closed status • • •	

Revised contents list for the data store SALE CONTRACT and the new contents list for the reference data store BUYERS:

Data structure: Sale contracts

Group	Repetitions	Components	Comments
		Sale contract number Buyer code Commodity Sale contract price Open/closed status • • •	

Data structure: Buyers

Group	Repetitions	Components	Comments
		Buyer code {Individual name Company name} Address City State Zip code	

Figure 7-2 Revisions to the preliminary contents lists that will accommodate reference data stores

given commodity to be grouped. But if those data structures contained similar, yet nonidentical, commodity references, they could not be grouped accurately.

To implement this standardization, I added a reference data store to the system called COMMODITY. Then, when application programs are developed, they'll be designed so they access the COMMODITY data store to validate operator-entered codes. By providing this kind of standardization, the value of the commodity data element in all data stores will be uniform. Then, it can be used reliably for summary and grouping functions.

After you've identified all of the reference data stores you think are appropriate, you need to add modules to the system structure chart for maintaining these data stores. Figure 7-3 shows how I handled this. First, I added a second-level control module to the system structure chart called MAINTAIN BROKERAGE REFERENCE FILES. Then, subordinate to it, I added modules to maintain each of the reference files I identified. I also added modules to list the contents of each of the reference data stores. Listings allow the user to see what codes are assigned to each occurrence of the structure and can facilitate maintenance. Last, I grouped the maintenance and listing modules by data store and put them under a controlling module that allows the selection of the maintenance or listing function.

Identifying archive data stores   Not only did I omit reference data stores from the model DFD, but I also omitted other classes of data stores. Specifically, I omitted (1) data stores that contain archive data used only in extractions, (2) control data stores, and (3) data stores internal to processes. Since control and internal data stores are specific to processes, their design can and should be deferred until you specify the processes that use them. But you should identify *archive data stores* at this time.

To identify archive data stores, you need to examine all of the extraction processes on the system structure chart. As you examine them, look for groupings of data elements that are required but that aren't part of the original model DFD. As you identify those groups, evaluate whether or not they represent groupings of data that need to be maintained by the

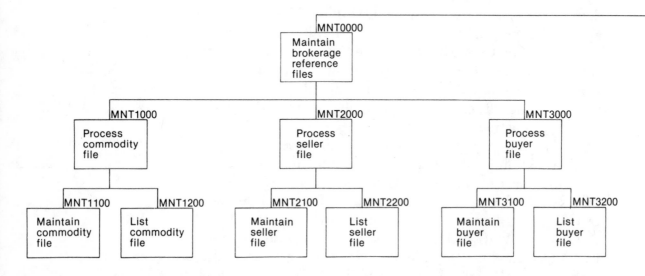

Figure 7-3   The system chart for the brokerage system showing the addition of the
maintain-brokerage-reference-files leg

system or that can be accumulated from other data stores on
demand. To do this, you need to have a thorough understanding of the storage and performance limitations of your system.
Without belaboring the thought processes I went through to
identify them, I decided that the brokerage system needed archive data stores for settlements, invoices, payables activity
history, and receivables activity history. When the system is
implemented, all four will contain information that's required
by extraction processes.

When you decide that a data grouping should be implemented as an archive data store, create a contents list for
it. Typically, an archive data store is created and maintained
by the operational modules of a system. As a result, you don't
have to design new modules to maintain them. They will
already appear on the system structure chart.

Even though you're finished with the model DFD as far
as developing the system structure chart is concerned, I
recommend that you add the archive data stores to it. As you
do, make the appropriate connections of the new archive data

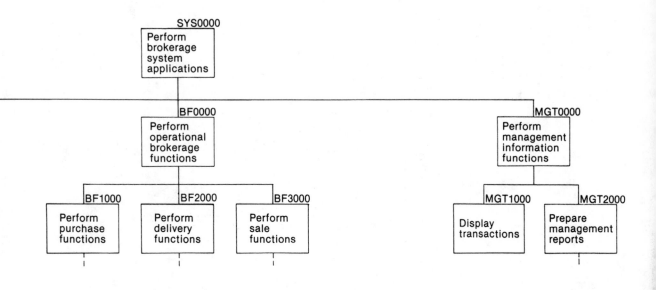

stores to their related processes. Although you won't use the
model DFD for anything other than documentation after you
do this, I find that this helps me understand the relationships
between the archive data stores and the other data stores and
processes. In figure 7-4, you can see the brokerage system
DFD after I added the four archive data stores to it.

As an aside, I didn't bother to add reference data stores
to the model DFD. Since they typically are connected to
many processes, it's difficult to add them to the DFD without
making a confusing mess. In addition, adding reference data
stores to the DFD would contribute little. After all, the
operational processes of the DFD don't maintain reference
data stores.

Step 2: Draw a data access diagram to show required access paths

One of the most critical tasks in designing a database is determining how the information in the system's data stores should

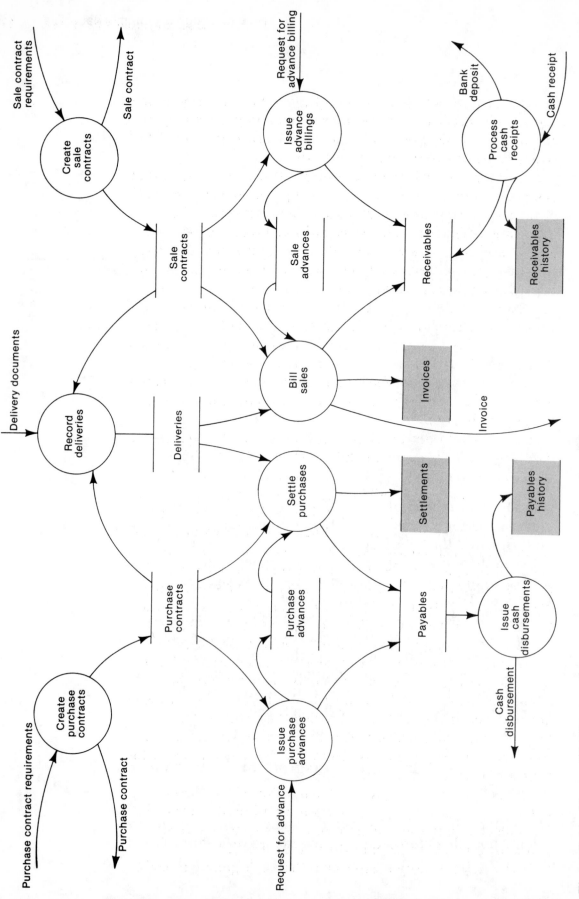

Figure 7-4 The model DFD for the brokerage system showing the addition of the archive data stores

be retrieved, or accessed. To do this, you must identify all the ways a user needs to retrieve information. So you start by examining all of the modules of the system structure chart. This is closely related to process specification.

To keep track of required access paths, I recommend that you create a *data access diagram* as you examine each module of the system structure chart. A data access diagram shows the relationships between access paths and data stores. Your goal should be to draw a single data access diagram that has only one occurrence of each data store and access path, with their relationships shown by connecting lines.

To develop a single data access diagram for all data stores from scratch is difficult. Instead of doing this, I suggest you begin by drawing a separate data access diagram for each data store. That means you'll duplicate some access paths. Then, when you've completed all of the individual data access diagrams, it's relatively easy to combine them into a single, integrated diagram.

To illustrate this development process, think about the data requirements of the module ENTER DELIVERIES (BF2100) in the structure chart in part 6 of figure 6-8. By referring to the model DFD in figure 7-4, you can see that it will draw data from the contract data stores and will add data to the delivery data store. What must happen in BF2100 is that a purchase contract and a sale contract for the same commodity code must be matched. A single contract must be uniquely identifiable by a contract number. As a result, the PURCHASE CONTRACT NUMBER is a unique access path to PURCHASE CONTRACTS, and the SALE CONTRACT NUMBER is a unique access path to SALE CONTRACTS.

Figure 7-5 shows how I represent unique access using a data access diagram. A box represents the data store, and a circle represents the data element that allows access to it. An arrow connects the circle and the box. In this case, the arrow has only one head, which means that one occurrence of the contract number provides access to a unique occurrence of the contract structure in the data store.

Figure 7-5 is so simple, it's trivial. But, it's a starting point for representing all access paths to all data stores and for understanding their relationships. As I continue to examine BF2100, I realize that the module ENTER DELIVERIES is more complicated than I originally thought.

Figure 7-5   A simple data access diagram showing unique accesses to the data stores PURCHASE CONTRACTS and SALE CONTRACTS

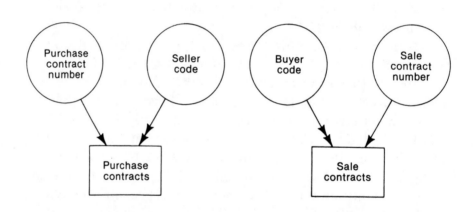

Figure 7-6   Complete data access diagrams for the data stores PURCHASE CONTRACTS and SALE CONTRACTS

Sometimes the users will know the proper contract numbers when recording a delivery. But most of the time, they'll only know from what seller the delivery came and to what buyer it went. As a result, the module must let the user select the right contracts for a buyer and a seller from all existing contracts. Accordingly, the two contract data stores must be accessible not only by contract number, but by seller and buyer as well. To show this, I added another access path to the diagrams for the two contracting data stores.

Figure 7-6 shows the revised data access diagrams. Notice here that the arrows connecting the seller and buyer codes to the data stores are double-headed. This indicates that a single buyer or seller code may be associated with multiple occurrences of the data stores' structures.

To complete the data access diagrams, you must continue this process for all modules of the system structure chart. Although this may seem to be an exhausting process, it must be done. In most cases, though, after you've examined a few modules, you'll see that the same access paths occur over and over again.

Figure 7-7 shows the set of diagrams I had after I examined all of the modules of the brokerage system's structure chart. Now that you understand the symbols I use in the diagrams, you should be able to understand all of the required access paths to all of the system's data stores just by looking at the diagrams. Notice that for most of the data stores, there is a way to access a unique occurrence of the data structure (that is, a single-headed arrow points to the box representing the data store). In many cases, there's also a required access that's one-to-many. In three cases, unique access is not required, but one-to-many access is.

After you've completed an individual data access diagram for each data store, your next task is to combine them into a single, integrated diagram as I did in figure 7-8. What I did here was combine all of the independent diagrams of figure 7-7 into one where no access-path circle or data-store box appears more than once.

As you develop an integrated data access diagram, you'll find that it's difficult, if not impossible, to keep from crossing lines. That just suggests that the access paths you've identified relate to several or many data stores. If you find yourself with several data stores with exactly the same access paths to each, that suggests that you can consider combining them into one data store. In this case, the data stores SETTLEMENT, PURCHASE ADVANCE, and PAYABLES are all accessible through identical keys. (The same is true for the corresponding data stores related to sales.) For the final database design, then, you need to decide whether these will be implemented as separate data stores or as a compound data store. In figure 7-8, I've drawn a large rectangle around the data stores that are accessed using the same keys. This is a visual reminder that they are related and may be combined in the final database design.

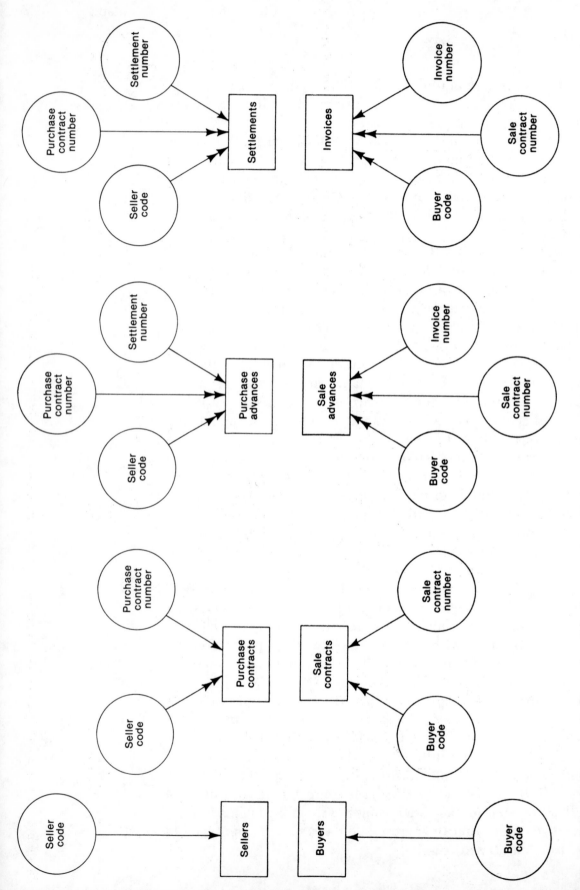

Figure 7-7 Data access diagrams for all brokerage system data stores identified thus far (part 1 of 2)

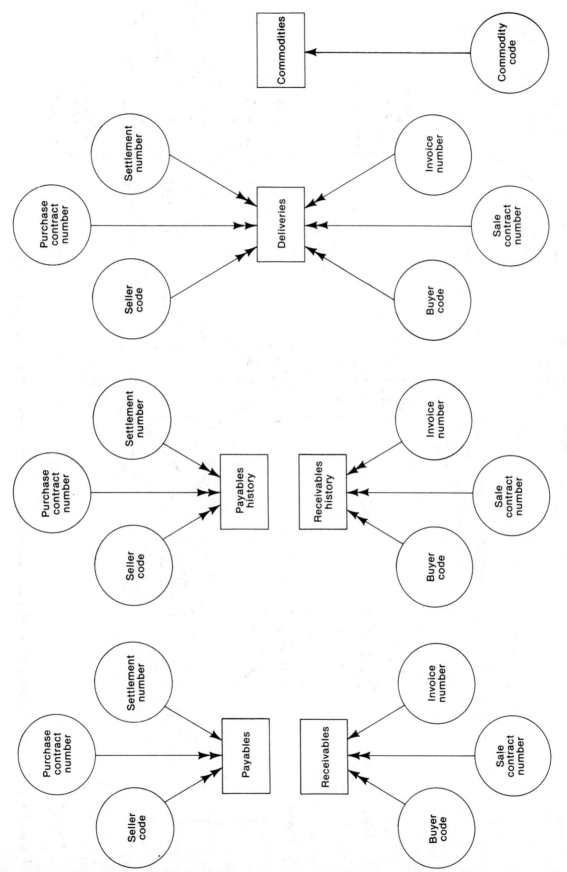

Figure 7-7 Data access diagrams for all brokerage system data stores identified thus far (part 2 of 2)

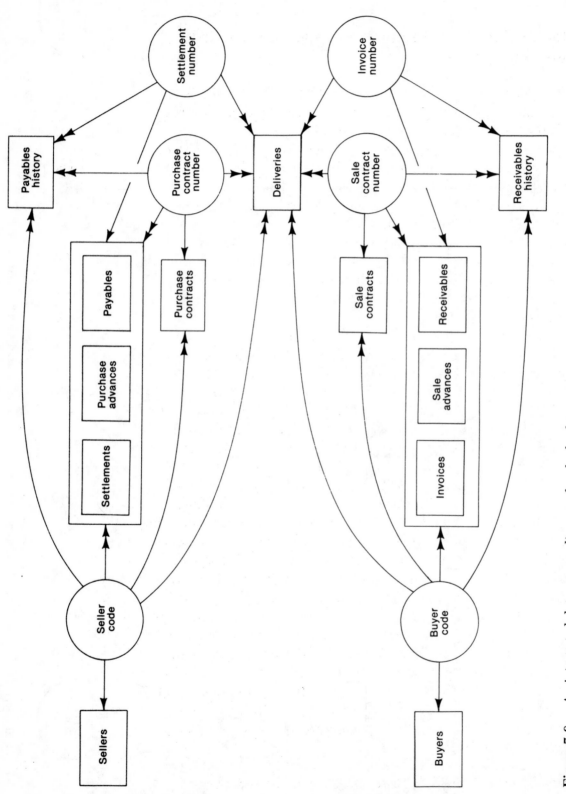

Figure 7-8 An integrated data access diagram for the brokerage system

Step 3: Draw a data hierarchy diagram to clarify data relationships

After you've identified the access paths for all data stores, I recommend that you develop a *data hierarchy diagram* that reflects the hierarchical relationships between those data stores. In most systems, there is a logical hierarchy among the data stores, and recognizing it will help you understand how to implement the database.

A systematic way to develop a hierarchy diagram is to work from the integrated data access diagram. To start, examine each access path that allows one-to-many access to a data store. If any of those access paths also provides *unique* access to a second data store, then you can assume that the second data store is superior to the first in the hierarchy of data stores.

For instance, look in figure 7-8 at the access path SELLER CODE. SELLER CODE provides one-to-many access to several data stores (PURCHASE CONTRACTS, SETTLEMENTS, PURCHASE ADVANCES, PAYABLES, DELIVERIES, and PAYABLES HISTORY). It also provides unique access to SELLERS. The conclusion is that the data store SELLERS is superior to PURCHASE CONTRACTS and the others I just listed.

That's one way to recognize hierarchical relationships. Another is through reasoning and common sense. In short, look for obvious owner/member and parent/child relationships and correspondences. If, for example, you know that a particular seller may have many contracts and that a contract cannot exist unless it's related to a seller, this establishes the hierarchy. In this case, the contracts are subordinate to the seller.

Figure 7-9 is a data hierarchy diagram that shows the relationship between SELLERS and PURCHASE CONTRACTS. As in a data access diagram, the boxes represent data stores, and a double-headed arrow indicates a one-to-many relationship. By identifying other hierarchical relationships, it's possible to develop a chart that includes most, if not all, of a system's data stores.

Figure 7-10 shows the data hierarchy diagram for the brokerage system's data stores. In it, you can see a clear

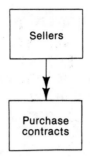

Figure 7-9   A simple data hierarchy diagram

hierarchical structure. Notice the relationship of PAYABLES to PURCHASE ADVANCES and SETTLEMENTS. One occurrence of PAYABLES exists for each occurrence of PURCHASE ADVANCES or of SETTLEMENTS. That's what the single-headed arrow indicates. (The relationships are the same on the sale side of the diagram.)

I would like to point out that these diagrams are not DL/I hierarchy charts, although they resemble them. These data hierarchy diagrams are just intended to help you gain a clearer understanding of a system's database requirements.

Step 4: Evaluate and refine the database plan

As you continue to work on process specification, you'll come to realizations that will affect the system's database structure. For example, as I examined the relationship between PURCHASE ADVANCES and SETTLEMENTS (and the one between SALE ADVANCES and INVOICES), I realized that two new data stores were required. Let me explain why for the purchase side of the diagram. The logic applies as well for the sale side.

It's possible that an unlimited number of advances can be applied against a single final settlement. Similarly, a single advance can be applied to an unlimited number of final settlements. To keep track of this relationship between advances and settlements, I added a new data store that I called PURCHASE ADVANCE APPLICATIONS as shown in the

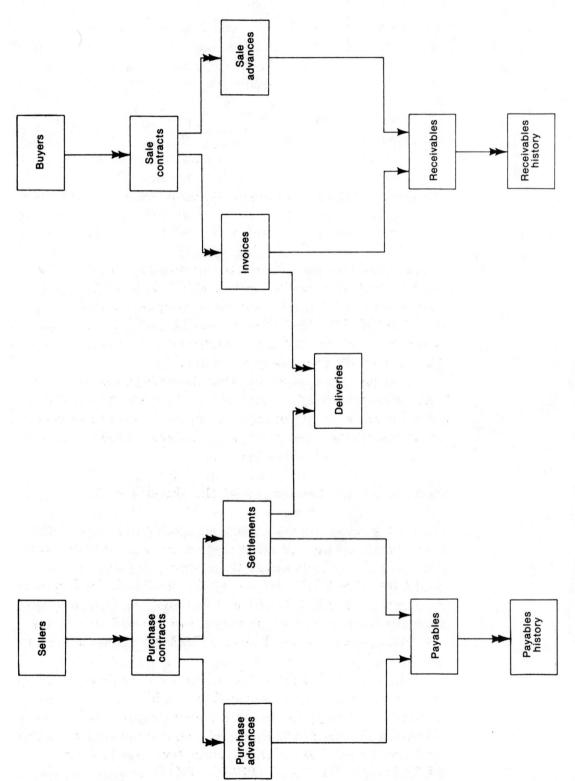

Figure 7-10   The first data hierarchy diagram for the brokerage system

diagram in figure 7-11. One occurrence of its structure
records the relationship between one advance and one final
settlement. If a single settlement has five advances applied
against it, then five occurrences of the data structure will be
created.

I decided to create a separate data store rather than add
data elements to the advance and settlement data structures
for two reasons. First, in most cases, settlements won't have
any advances applied to them. If I provided space for recording detailed information for many advances in the settlement
data store, I'd waste some space and sacrifice some processing
efficiency. Second, the number of possible applications on
either side is unlimited. Although the number is usually no
greater than three, I couldn't be sure that any limit I'd set for
a repeating data structure within the settlement or advance
structure wouldn't be exceeded at some time. (I'll have more
to say about record design and repeating groups later in this
chapter.)

Figure 7-11 shows the system data hierarchy diagram
with the advance application data stores added to it. You can
see that the purchase application data store can be accessed
either through a purchase advance or a settlement and, in
both cases, the relationships are one-to-many. The situation is
the same on the sale side of the diagram.

As I continued to evaluate the data hierarchy diagram, I
recognized a problem related to the data store DELIVERIES.
Because a delivery is not automatically settled or invoiced
when it's entered, the data hierarchy diagrams in figure 7-10
and 7-11 are misleading. An unsettled delivery cannot be
associated with a settlement, although it will be associated
with a specific purchase contract. Similarly, an uninvoiced
delivery cannot be associated with an invoice, although it is
associated with a specific sale contract.

This presents a kind of logical problem, and its solution
will, to a great extent, be dictated by the capabilities and
constraints of the system on which the database is implemented. On the system I used for this application, however,
I was able to implement this structure *and* allow for deliveries
in the delivery data store that didn't relate back to a settlement or an invoice. I did this by assigning unsettled deliveries
a settlement number of zero and unbilled deliveries an invoice

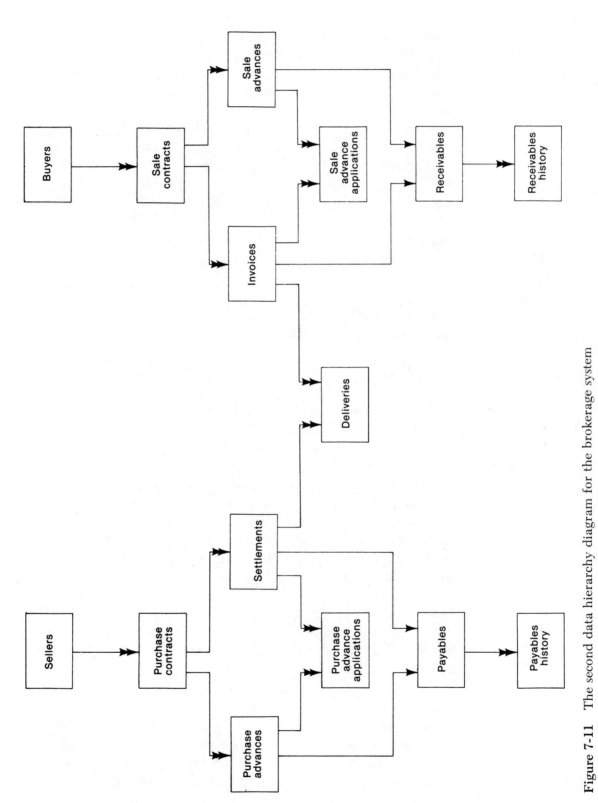

Figure 7-11   The second data hierarchy diagram for the brokerage system

number of zero. That way, such deliveries can exist along with those that have been completed (both billed and settled). They don't fit into the hierarchy logically, however, because there is no settlement number zero and no invoice number zero.

As I've said, this type of detailed design work demands creativity, insight, and experience, plus detailed knowledge of your computer system. So I can only give you some rough guidelines for evaluating what you've done thus far. As a final test of your database plan, consider it from the point of view of the user and the application. Is it logical from that perspective? If it is, you're on the right track. If it isn't, your design needs work.

Step 5: Decide how to package the system's data elements

By now, you've made a contents list for each data store, you've identified the paths by which each store will be accessed, and you've defined the hierarchical relationships among the stores. After you've gone this far in the database design process, the packaging and documentation work that remains is fairly straightforward.

If you're developing a system in a large company using database software (a DBMS), you'll probably relinquish database design to the database administration (DBA) group at this time. A DBA group is responsible for the overall operation of the database system and must balance the needs of all subsystems, of which yours will be only one. The DBA group will develop the physical implementation of your database requirements and will prepare the documentation you'll need for your program specifications (such as application database views). If this is your situation, the rest of the material in this chapter won't be of much use to you.

If you work in a small shop that uses a DBMS, you may be your own database administrator. In that case, you'll need to have an in-depth understanding of the DBMS installed on your system. Although I'm not going to present the details of database design for a DBMS in this chapter or book, I encourage you to read on anyway. In a few moments, I'll give you some guidelines for formatting the record descriptions that you pass along to your programming staff.

If you work in a more traditional file-based environment, the packaging decisions that determine the physical implementation of your database will be your responsibility. At this time, then, I'd like to give you some ideas about database design in a file-based environment.

For the most part, the files of your system will correspond to the data stores you've already identified. In most cases, though, you will make some packaging decisions related to these files. These decisions will be based on efficiency and operational considerations. One important packaging decision you'll have to make is which access paths you'll implement to the system's data stores.

When you provide one unique access path to a data store, that's straightforward enough. You normally implement it as an indexed file with the file's key as the access path. But if a file must be accessed along a one-to-many access path or along more than one unique access path, then matters are complicated somewhat.

Using alternate indexing   If your system provides for *alternate indexing* (also called *alternate keying*), you can easily implement both multiple-unique and one-to-many access paths. Basically, when a file is created with an *alternate key*, a separate index is kept for that key in addition to the one for the file's *primary key*. Although a file's primary key must contain a unique value for each record in the file, its alternate key or keys need not. As an alternate indexed file is changed through additions, deletions, or modifications, all indexes (primary and alternate) are updated to keep all access paths to the file current. The effect of alternate indexes is that a single data file can appear to be presorted along each alternate key path. By specifying at what key value to begin retrieving records along a particular path, one-to-many access is easy and efficient.

Before you decide that a file should have an alternate index, however, you should carefully evaluate the implications of the decision. Of course, it's desirable to have multiple access paths for a file. But, if accesses along an alternate path are infrequent or don't have to be performed rapidly, you should consider other alternatives. In some circumstances, for example, adding an alternate key to a file can degrade performance considerably. So you must be aware of potential update and retrieval inefficiencies, of how often the alternate indexed file will need to be reorganized, and of what increases in disk storage the file will require.

To illustrate, look at figure 7-12. This is a summary of disk storage requirements for an alternate indexed file on a minicomputer system on which I've worked. This example is for a hypothetical employee file with 300-character records that have a 5-character employee number as the primary key. If the file is to contain 5,000 records and is created without alternate indexes, it will require 1.64 megabytes of disk storage (approximately 1.62 for the data records themselves and .02 for the primary key index structure).

Imagine, however, that the file is created with six alternate indexes: social security number (9 characters), employee last name (20 characters), zip code (9 characters), and codes to indicate in what branch and division the employee works and his or her position (5 characters each). I think you'll agree that these are all reasonable (if not essential) alternate access paths. But if I created this file with all six alternate indexes, it would require 2.47 megabytes of disk storage. That's a space increase of 50 percent over what the file with only a primary key would require. You'll notice in figure 7-12 that the larger the alternate key is, the more space its alternate indexing structure requires.

To decide which (if any) of these six alternate key paths to implement on such a file, you need to consider how important each access path really is. For instance, access by last name may be essential for a variety of interactive functions and, as a result, probably should be an alternate key. On the other hand, access by zip code may be desirable only for batch jobs (for example, to prepare a mass mailing to all employees). In that case, it would probably be more reasonable not to implement zip code as an alternate key but, instead, to sort the file into zip code sequence when necessary.

When you decide whether or not to use alternate key paths, you've got to use your own judgement. But to some extent, you can rely on the knowledge and experience of your

NUMBER OF RECORDS ****5000

	DATA RECORD	LENGTH	LOW LEVEL	UPPER LEVELS	TOTAL BLOCKS	MEGABYTES
Employee record	DATA RECORD	*300	834		834	1.62891
Employee number	PRIMARY KEY	**5		8	8	.01563
Social security number	ALT KEY # 1	**9	70	1	71	.13867
Last name	ALT KEY # 2	*20	125	4	129	.25195
Zip code	ALT KEY # 3	**9	70	1	71	.13867
Branch code	ALT KEY # 4	**5	50	1	51	.09961
Department code	ALT KEY # 5	**5	50	1	51	.09961
Position code	ALT KEY # 6	**5	50	1	51	.09961
	ALT KEY # 7	***	0	0	0	0.00000
	ALT KEY # 8	***	0	0	0	0.00000
	ALT KEY # 9	***	0	0	0	0.00000
	ALT KEY #10	***	0	0	0	0.00000
	ALT KEY #11	***	0	0	0	0.00000
	ALT KEY #12	***	0	0	0	0.00000
	ALT KEY #13	***	0	0	0	0.00000
	ALT KEY #14	***	0	0	0	0.00000
	ALT KEY #15	***	0	0	0	0.00000
	ALT KEY #16	***	0	0	0	0.00000
	ALTERNATE INDEXED FILES REQUIRE A 1 BLOCK OVERHEAD				1	.00195
	REORGANIZED, UNCOMPRESSED FILE SIZE				1267	2.47461

Figure 7-12 Disk storage requirements for an indexed file with alternate keys

co-workers. Chances are that a problem you're wrestling with (or a similar one) has already been encountered by other people in your group. If so, you can learn from their experiences.

Implementing multiple-unique accesses without alternate indexing   Infrequently, a file may need two or more unique access paths. Imagine, for example, an employee file that must be accessed by employee number or social security number, both unique for an individual. Then, if your system doesn't support alternate indexing, you'll have to go to some trouble and introduce some inefficiencies into your database design to implement multiple-unique access paths. (I once worked on a small system that didn't support alternate indexes, but did allow multiple primary key structures for the same file. But that's the exception, and it was grossly inefficient to boot.)

If you have to implement more than one unique access path to a file, you can build your own *secondary index file*. A secondary index file allows you to access records in a file by a key other than the file's primary key. In this case, there's a one-to-one correspondence between the records in the secondary index file and the data file. Each record in the secondary index file contains two data elements: the two keys that are the unique access paths to the data file. The primary key of the secondary index file is the secondary, or alternate key, of the data file. To access the data file along the second unique access path, the proper record in the secondary index file is read. After the primary key value is extracted from the secondary index record, the appropriate record in the data file is read.

Implementing one-to-many accesses without alternate indexing   Typically, a set of related records in a one-to-many relationship is retrieved by reading all the records in a file and selecting those desired for further processing. This kind of processing can be done either with or without a preliminary sort. As you can imagine, this can result in tremendous inefficiencies in functions that require frequent retrievals of sets of records in one-to-many relationships (particularly on-line inquiry functions).

Another option, though, is to maintain summary information in a separate file that represents data from many related records. This approach can be effective for maintaining certain types of management information. For example, in an open-item accounts-receivable system, one open-item record is maintained for each unpaid invoice. To determine the total balance a specific customer owes, then, requires accessing all open-item records for that customer (a one-to-many relationship). But if the total balance due for each customer is maintained separately from the open-item records, it can be accessed directly (and rapidly) to provide the customer's current balance due.

For problems that do not lend themselves to summarization and for on-line functions that demand more rapid access than can be accomplished when an entire file has to be searched, there are other options. Among them are variations on the secondary index file concept I just described and linked-list techniques. To one degree or another, though, all are inefficient and complicated. As a result, I don't encourage you to use them unless you have no alternatives.

Handling repeating groups   Sometimes, it's hard to decide whether you should include a repeating data structure in a record for a file. The best physical implementation may depend on a variety of factors. Yes, one objective of database design for a system that uses a DBMS is to eliminate repeating groups (this is one of the goals of the process called normalization that leads to an optimal database design). But in a file-based system, eliminating repeating groups may *not* be desirable.

The major consideration when designing a record with a repeating group is deciding how many repetitions is the right number. If you pick a number that's too small, you'll face reprogramming problems when you encounter a situation where all positions are used and you need still more. The more repetitions you allow, of course, the less significant this problem becomes. Nevertheless, the risk of exceeding the maximum number of repetitions always exists unless an unalterable cut-off point exists. On the other hand, the more repetitions you allow, the more you waste disk space and processing resources.

An alternative to repeating groups is to create a separate record for each occurrence of a repeating data element. Often, this means repeating some standard data (key fields, for instance) in each record for a related group. And sometimes this is more wasteful than creating the record with a repeating group in the first place.

For some systems, the nature of the application dictates how many repeating groups to use. When that's the case, you're lucky. In the brokerage system, for example, the users told me that for a specific delivery, there would *never* be more than ten charges (or premiums) for quality variations. As a result, I felt confident about implementing the delivery record with a repeating group for charges. However, if the users had told me that there could be an unlimited number of charges related to a delivery, using a repeating group in the delivery record would have been inappropriate.

Restructuring data stores   In most cases, each of the data stores you've identified so far will become a file. However, in some instances it may be practical to combine two or more data stores into one file. It may also be desirable to break one data store down into two or more files. These are *restructuring* decisions.

If two or more data stores have identical access paths, that's an indication that they might be combined into one data store. But you need to use judgement here. If two data stores have the same access paths but radically different purposes, then they shouldn't be combined. An important distinction to make when deciding whether to combine data stores is the distinction between archive data and operational data. As a general rule, it's desirable for the two types of data to be kept in separate files.

If you look back to figure 7-8, you'll recall that, on the purchase side of the database, the data stores SETTLEMENTS, PURCHASE ADVANCES, and PAYABLES have identical access paths. Now it's time to decide if these data stores should be implemented as one, two, or three files.

The way I approached this problem is instructive. SETTLEMENTS and PURCHASE ADVANCES both represent original transactions that result in the creation of a payable item. They differ only in that SETTLEMENTS are transactions based on actual deliveries, and PURCHASE ADVANCES are transactions based on anticipated deliveries. Both contain similar data elements, and both are essentially archive data stores. After transactions are added to these data stores, they will not be changed. In contrast, PAYABLES is not an archive data store, but an operational data store that will be quite active. As a result, it makes good sense to separate it from the archived information in the other two data stores.

The decision? I implemented the three data stores as three separate files. From what I've explained so far, it seems entirely reasonable to keep PAYABLES separate and to combine SETTLEMENTS and PURCHASE ADVANCES because they're archive data stores with similar data elements. Actually, that *would* work. However, the PURCHASE ADVANCES data store has an operational element to it that makes it most efficient to keep it a separate file. As I developed the process specifications for the program that will create final settlements, I realized that it needs to access all purchase advances for a specific contract to determine which (if any) have an unapplied balance. But if advances and final settlements were mixed together, the program would have to read through both types of transactions to select out the advances. Since settlements outnumber advances by a factor of about 50 to 1, this would be quite inefficient. (As you'd expect, we packaged the corresponding data stores on the sale side of the database in the same way.)

When you restructure data stores, you must also consider what information needs to be maintained and for how long. The delivery data store in the brokerage system illustrates what I mean. The charge data in the delivery record (which can have up to ten occurrences) is used only during settlement and invoicing. After a delivery has been completely processed and the charge data has been printed on the related settlement and invoice, charge data is no longer required, not even for information retrieval. In this case, because the charge data would account for a large percentage of the space each delivery record requires and because the delivery history file will eventually contain many thousands of records, I decided to implement the delivery data store not as one, but as three separate files.

The first of these files is the permanent delivery file (which I called DELIVERY). It will contain one record for each delivery recorded and will be accessible by seller code, buyer code, purchase contract number, sale contract number, settlement number, and invoice number. Each record will contain weight, delivery, and summary pricing information (not detailed charge data). The second of these files will contain only unsettled deliveries (called PURCHDEL), including the charge data. Its records will contain all of the data necessary to prepare a settlement. The third file corresponds to PURCHDEL on the sale side of the database. It's called SALEDEL.

When a delivery is recorded, then, three records will be created: one each in DELIVERY, PURCHDEL, and SALEDEL. When that delivery is settled, the file DELIVERY will be updated so its record is accessible by settlement number, and its record in PURCHDEL will be deleted. Similarly, when it's invoiced, DELIVERY will be updated so its record will be accessible by invoice number, and its record in SALEDEL will be deleted.

This implementation will do more than save valuable disk space. It also will make the settlement and invoicing functions more efficient. Instead of working on a file that contains 50,000 records, those programs can process much smaller files (say files of 200 to 300 records) that contain only unsettled or unbilled deliveries.

I hope these examples have impressed upon you the close connection between process specification and database design. Also, I hope they've given you some insight into how many factors need to be considered when making database decisions.

Step 6: Document the physical database design

After you've determined what the physical organization of your system's database will be, your last step is to document that organization. You'll need to create record layouts and other supporting materials that will be available when the application programs are developed. Fortunately, this documentation task is mechanical.

First, you'll need to decide what names your system's files will have. Try to assign file names that are as meaningful as possible, but that meet your system's naming requirements and your shop's standards. It can be difficult to create an eight-character name that meaningfully describes a file, but do your best. Almost always, you'll be forced to use abbreviations. For example, in the brokerage system, I named the PURCHASE CONTRACTS file PURCHCON. Although it's an abbreviation, it's still meaningful.

To develop record layouts, work directly from the contents lists you've already developed for the system's data stores. Naturally, if you've decided to combine or divide data stores, you'll need to make the appropriate adjustments as you work from the contents lists to the record descriptions. In most cases, however, all you need to do to create a record description is translate the free-form contents list into whatever programming language you'll be using to develop the system.

As you create the detailed record descriptions, you'll have to make decisions about each field's length and characteristics. Where possible, store numeric data in a form that's efficient for computations. Be sure all fields are the proper size. If similar fields exist in several files, be sure they're compatible in size and data representation. Also, be aware of blocking factors that can affect program performance as you organize records. On some systems, there are restrictions on where in a record a key or a potential sort field may be positioned (for example, in the first 256 characters of a large record). Obviously, you need to be aware of any such restrictions that apply in your situation.

And here's a strictly pragmatic suggestion that comes from experience. Position identifying data (keys, names, important codes, and so on) as close to the beginning of a record as you can. If you do, it will be easier for you to identify records if you need to examine the contents of the file later on.

Because the database ties the various development functions together, it's important that all developers see the database in the same way. To insure uniformity, then, you should create standard code for record descriptions and other file-related descriptions that can be copied into the applica-

tion programs that make up the new system. Not only does
the use of standard code help insure that all programmers are
using the same descriptions, it also cuts down on coding time.
Since I developed the brokerage system in COBOL, my standard descriptions are in the form of COBOL COPY members.
If you work in PL/I, you'll use "included text." And if you
work in some other language, you should either have this
capability or simulate it.

If you look at figure 7-13, you'll see the COPY member I
created for the record description of the file named
DELIVERY. You can see that it's formatted in a rigid way
and follows some coding conventions. I'll have more to say
about coding conventions in chapter 10. For now, you should
realize that one of the most important goals of program
development is to create programs that are easy to understand. The easier a program is to understand, the easier it is
to test and maintain.

To make record descriptions more readable, I group
logically related fields, use subordination to show the relationships between fields, and indent elements to stress those relationships. Also, the names I assign to fields are as descriptive
as I can make them.

In figure 7-13, you can see that all of the data names
begin with a prefix: DR. These prefixes make it easier to identify what record a particular data element is part of. As a
result, they contribute to program readability. I use a prefix
that's an acronym created from the record name in all record
descriptions I create.

When similar data elements exist in several files, use consistency when naming them. Keys, for example, should have
the same name but should be uniquely identified by prefixes
that identify the structures they're part of. For instance, the
seller code in the brokerage system appears in several data
structures: purchase contracts, settlements, payables, and
deliveries, among others. So I assigned these names to the
fields:

SR-SELLER-CODE	SELLERS
PCR-SELLER-CODE	PURCHASE CONTRACTS
PSR-SELLER-CODE	SETTLEMENTS
APR-SELLER-CODE	PAYABLES
DR-SELLER-CODE	DELIVERIES

```
01    DELIVERY-RECORD.                                              DELIVERY
*                                                                   DELIVERY
      05   DR-DELIVERY-KEY                     PIC 9(7).            DELIVERY
      05   DR-COMMODITY-CODE                   PIC X(5).            DELIVERY
      05   DR-DELIVERY-DATE.                                        DELIVERY
           10   DR-DELIVERY-YEAR               PIC 99.             DELIVERY
           10   DR-DELIVERY-MONTH              PIC 99.             DELIVERY
           10   DR-DELIVERY-DAY                PIC 99.             DELIVERY
      05   DR-ENTRY-DATE.                                           DELIVERY
           10   DR-ENTRY-YEAR                  PIC 99.             DELIVERY
           10   DR-ENTRY-MONTH                 PIC 99.             DELIVERY
           10   DR-ENTRY-DAY                   PIC 99.             DELIVERY
      05   DR-CARRIER-ABBREVIATION             PIC X(10).          DELIVERY
      05   DR-CARRIER-BILL-OF-LADING           PIC X(10).          DELIVERY
*                                                                   DELIVERY
      05   DR-SALE-DATA.                                            DELIVERY
           10   DR-BUYER-CODE                  PIC X(16).          DELIVERY
           10   DR-SALE-CONTRACT.                                   DELIVERY
                15   DR-SALE-CONTRACT-NUMBER   PIC 9(5).           DELIVERY
                15   DR-SALE-CONTRACT-SEQUENCE PIC 9.              DELIVERY
           10   DR-INVOICE-NUMBER              PIC 9(5).           DELIVERY
           10   DR-SALE-AMOUNTS                USAGE IS COMP.      DELIVERY
                15   DR-SALE-GROSS-PRICE       PIC S9(5)V99.       DELIVERY
                15   DR-SALE-CHARGES           PIC S9(5)V99.       DELIVERY
                15   DR-SALE-FREIGHT           PIC S9(5)V99.       DELIVERY
           10   DR-SALE-WEIGHT-CERTIFICATE     PIC X(15).          DELIVERY
           10   DR-SALE-GRADE-CERTIFICATE      PIC X(15).          DELIVERY
           10   DR-SALE-WEIGHTS                USAGE IS COMP.      DELIVERY
                15   DR-SALE-GROSS-WT-POUNDS   PIC S9(9).          DELIVERY
                15   DR-SALE-TARE-WT-POUNDS    PIC S9(9).          DELIVERY
                15   DR-SALE-DOCKAGE-WT-POUNDS PIC S9(9).          DELIVERY
                15   DR-SALE-DOCKAGE-PERCENTAGE PIC S99V9.         DELIVERY
*                                                                   DELIVERY
      05   DR-PURCHASE-DATA.                                        DELIVERY
           10   DR-SELLER-CODE                 PIC X(16).          DELIVERY
           10   DR-PURCH-CONTRACT.                                  DELIVERY
                15   DR-PURCH-CONTRACT-NUMBER  PIC 9(5).           DELIVERY
                15   DR-PURCH-CONTRACT-SEQUENCE PIC 9.             DELIVERY
           10   DR-SETTLEMENT-NUMBER           PIC 9(5).           DELIVERY
           10   DR-PURCH-AMOUNTS               USAGE IS COMP.      DELIVERY
                15   DR-PURCH-GROSS-PRICE      PIC S9(5)V99.       DELIVERY
                15   DR-PURCH-CHARGES          PIC S9(5)V99.       DELIVERY
                15   DR-PURCH-FREIGHT          PIC S9(5)V99.       DELIVERY
           10   DR-PURCH-WEIGHT-CERTIFICATE    PIC X(15).          DELIVERY
           10   DR-PURCH-GRADE-CERTIFICATE     PIC X(15).          DELIVERY
           10   DR-PURCH-WEIGHTS               USAGE IS COMP.      DELIVERY
                15   DR-PURCH-GROSS-WT-POUNDS  PIC S9(9).          DELIVERY
                15   DR-PURCH-TARE-WT-POUNDS   PIC S9(9).          DELIVERY
                15   DR-PURCH-DOCKAGE-WT-POUNDS PIC S9(9).         DELIVERY
                15   DR-PURCH-DOCKAGE-PERCENTAGE PIC S99V9.        DELIVERY
```

Figure 7-13   A record description for the file DELIVERY presented as a COBOL COPY member

```
SELECT DELIVERY ASSIGN TO "DELIVERY"  "DISK"                        DELIVERY
                ORGANIZATION IS INDEXED                             DELIVERY
                ACCESS IS DYNAMIC                                   DELIVERY
                RECORD KEY IS DR-DELIVERY-KEY                       DELIVERY
                ALTERNATE RECORD KEY                                DELIVERY
                1  DR-BUYER-CODE          WITH DUPLICATES           DELIVERY
                2  DR-SALE-CONTRACT       WITH DUPLICATES           DELIVERY
                3  DR-INVOICE-NUMBER      WITH DUPLICATES           DELIVERY
                4  DR-SELLER-CODE         WITH DUPLICATES           DELIVERY
                5  DR-PURCH-CONTRACT      WITH DUPLICATES           DELIVERY
                6  DR-SETTLEMENT-NUMBER   WITH DUPLICATES.          DELIVERY
```

Figure 7-14   A COBOL SELECT statement for the file DELIVERY presented as a COPY member

If several blocks of text are required to identify a file completely, make them all standard code that's copied into the programs that use the files. For example, in file-oriented COBOL, the SELECT statement for a file should be a COPY member. Figure 7-14 shows my COPY member for the SELECT statement of the file named DELIVERY. If you use a DBMS, the descriptions of the database views used in application programs should be standard code as well. For example, if you work in COBOL and use an IBM DL/I DBMS, all PCBs should be COPY members.

Discussion

As you should realize by now, database design is a creative process. Since you need experience and insight to do it effectively, I can't give you a definitive method for doing it. To complicate matters, hardware and software systems vary widely from one shop to another. Nevertheless, I think the general approach I've explained in this chapter can give you a fresh approach to database design. If you try these techniques, I'm confident that they'll contribute to your design efforts.

Terminology

reference data store
archive data store
data access diagram
data hierarchy diagram
alternate indexing
alternate keying
alternate key
primary key
secondary index file
restructuring

Objectives

Given a model DFD, process requirements, and information about your installation's hardware and data-management software:

1. Identify all of a new system's data stores and the access paths to them.

2. Create data access and data hierarchy diagrams for the system's data stores.

3. Design an effective file and keying structure for the system.

4. Create useful database documentation for the development of the programs of the system.

Chapter 8

How to create program specifications

 In this chapter, I'll show you how to create practical specifications for the programs of a system. These specifications will be used by the programmers who develop the programs of the system. Essentially, all you need to develop as program specifications is a set of report and screen layouts (if a system module requires them) plus a program overview.

In some shops, you may not even need to create the report and screen layouts for the programs of a system. Sometimes, that's the responsibility of the programmer. Still, you'll need to establish a set of standards for report and screen formatting. In this chapter, then, I'll stress the importance of having a consistent approach to developing reports and screens.

A program overview is a simple document that lists each program's inputs and outputs and specifies what the program should do. In this chapter, my goal is to show you how to create overviews that are both useful and brief. In general, the briefer a program specification is, the less time you spend writing it and the less time a programmer spends trying to understand it. On the other hand, you can't sacrifice accuracy and completeness when you create program overviews.

Before you begin to create report and screen layouts and program overviews, you need to make some preliminary decisions. First, you need to decide if any program code can be standardized for use in more than one program. In other words, you must ask yourself what subprograms or segments of source code might be useful for your system. Second, you need to make final decisions about the packaging of system modules into programs. As a result, I'll explain these tasks first. Next, I'll present some ideas for report and screen design. Last, I'll show you how to create program overviews.

How to use standard programming code

Before you begin to specify programs, you should look for functions that are common to two or more programs. When you identify them, you should decide whether to implement them as subprograms that can be used throughout the system. Obviously, you should identify all useful subprograms *before* you begin program specification so you can refer to them in your program overviews.

You'll recall from the last chapter that using standard code for file and record descriptions contributes to programmer productivity by reducing coding. It also insures system integrity because all users have the same view of the database.

Using subprograms offers similar advantages. To illustrate, imagine a function that is common to many programs, such as editing a date to make sure its fields contain logical values. Although the function is simple, it still takes time to code and test the routine to do it. So if each programmer codes the routine whenever it is required in a program, programmer productivity is reduced. Not only is the same work done over and over again, but there's a chance that some of the date-edit routines in some programs won't work correctly. In contrast, a fully tested date-edit routine used as a subprogram can eliminate both of these problems.

In general, if you find a function that needs to be done more than once in a system, it should be implemented as a subprogram. Sometimes, of course, minor differences in the

---

Date- and time-handling functions:

    Convert a Gregorian date to a Julian date (and vice versa)

    Compute the elapsed time in days between two dates

    Determine the day of the week from the date

    Resequence the elements in a date

    Convert 24-hour time to a.m./p.m. time (and vice versa)

    Compute elapsed time

Editing functions:

    Validate a date

    Validate a state code

    Validate a zip code

Character-manipulation functions:

    Center a left-justified character string in a field

    Compress groups of spaces to a single space within a character string
        to create a new string that looks like a sentence

Output-related functions:

    Format a standard heading for a report

    Print a screen image

Figure 8-1   Some general subprogram functions

requirements from one program to another may rule this out. But, in most systems, there are more than a few functions that should be implemented as subprograms.

In our shop, we use about two dozen subprograms regularly. Naturally, some are used more frequently than others. Still, because we have them and because they are available to our entire programming staff, they have saved many hours of coding, testing, and maintenance.

Figure 8-1 lists some common functions that you should consider for your subprogram library. All are useful, and all are fairly simple. Without much trouble, you can probably add to this list.

In addition to general functions like those in figure 8-1, look carefully at your system for special functions that should be implemented as subprograms. For instance, in the brokerage system, I decided to implement a subprogram that displays all valid commodity codes. Since several programs in the system require that a commodity code be entered, this display function proved to be useful.

After a subprogram has been developed (and, needless to say, thoroughly tested), it should reside in a library that is accessible to all programmers. Additionally, each programmer should have documentation for all available subprograms that describes their functions and shows how to invoke them. Usually, a one-page description is enough. For instance, figure 8-2 shows acceptable documentation for a COBOL date-edit subprogram.

In some cases, you may want to implement standard code not as an executable module (a subprogram), but rather as source code that can be copied into programs where appropriate. If a programmer needs to do a function that is similar to another one that is available as source code, he can copy the source code into his program. Then, he can make minor modifications to it and save programming time. When you use segments of source code in this way, they should be available to the programmers in a development library. And each programmer should have documentation for each available segment of code.

How to package system modules into programs

For the most part, each work module on the system structure chart becomes a single program in the final system. In some cases, though, it is desirable to combine some work modules into one program or to divide one module into two or more programs. These combinations and divisions are referred to as *program packaging*. As you will see, packaging decisions are somewhat arbitrary. And to some extent, you should let the user determine how the functions of a system should be packaged.

When you package programs, you must be aware that hardware limitations can affect your decisions. In particular, the maximum program size allowed by your computer can limit the number of functions you can package into a single program. For instance, I once developed programs for a

---

Subprogram name:	DATEDIT
Programmer:	Steve Eckols
Date:	6-30-83

Function:
This program evaluates a six-character date in the format MMDDYY (month-day-year) and sets a one-character switch field to "Y" if the date is logically valid or to "N" if it is not.

Editing:
All three fields within the date must be numeric.

The month field must have one of these values:
01, 02, 03, 04, 05, 06, 07, 08, 09, 10, 11, or 12.

If the month field has value 02, then the day field value must be between 01 and 28 if it is not a leap year (i.e., if the year field value is not evenly divisible by 4).

If the month field has value 02, then the day field value must be between 01 and 29 if it is a leap year (i.e., if the year field value is evenly divisible by 4).

If the month field has value 01, 03, 05, 07, 08, 10, or 12, then the day field value must be between 01 and 31.

If the month field has value 04, 06, 09, or 11, then the day field value must be between 01 and 30.

If all of these editing criteria are met, the switch field value is set to "Y". Otherwise, the switch field value is set to "N".

Use:
To use this program, code these (or similarly named) fields in the WORKING-STORAGE SECTION of your COBOL program:

```
05   VALID-DATE-SW      PIC X.
     88   VALID-DATE              VALUE ''Y''.
          .
          .
          .
05   CURRENT-DATE.
     10   CURRENT-MONTH  PIC XX.
     10   CURRENT-DAY    PIC XX.
     10   CURRENT-YEAR   PIC XX.
```

Then, code this CALL statement in your PROCEDURE DIVISION:

```
CALL 'DATEDIT' USING VALID-DATE-SW
                     CURRENT-DATE.
```

Figure 8-2   Acceptable documentation for a subprogram

minicomputer that allowed a maximum of 60K for the object code of a COBOL program. Needless to say, that often forced me to divide a logical programming module into two or more smaller programs. So if you're working on a small system like a PC, you may be forced to make similar divisions, although this is less likely today than it was a few years ago.

In any event, I'm going to present packaging ideas assuming that you have no system limitations. So if you do, you must keep them in the back of your mind. Then, you may not be able to combine system modules as I suggest in this chapter. And you may be forced to divide programs due entirely to computer limitations.

If you do combine or divide modules on the system structure chart, you should make these changes to the chart. When you have packaged all the programs, then, your revised system structure chart will reflect the changes. As a result, you will be able to implement the structure chart with one application program for each of the work modules on the chart.

When to combine system modules   Generally, a system is easier to implement and maintain if each program provides for one, and only one, function. In some cases, though, some benefits result when you combine functions. Fortunately, then, when logical functions are clearly defined and when structured design techniques are used at the program level (a topic I'll cover in the next chapter), it's possible to combine several functions into a single program with little if any reduction in ease of development or maintenance.

A disadvantage to single-function programs is that they can cause operational inefficiencies in an interactive system. Although machine efficiency usually isn't that important in an interactive system, user efficiency is.

To illustrate, look back to the system structure chart for the brokerage system in figure 6-8. Specifically, look at the PERFORM PURCHASE CONTRACTING FUNCTIONS subsystem in part 2 of that figure. When implemented, an operator whose job is entering and maintaining contracts may, due to the nature of the source documents received, want to jump from one work module to another: creating contracts, fixing prices, making corrections, and changing statuses. But with these functions separate, as they are on the system structure chart, to do so requires leaving one program, passing up through the menu structure, and then starting a new program. Although that's not a major problem when only one level of menus is involved, it can be a real irritation if the operator has to pass through several levels.

Depending on programming restrictions and operator needs, then, it might be better to combine all of those functions into one module (and hence one program). In fact, from the programmer's perspective, this might be an efficient way to approach the problem. As you can imagine, many of the tasks that must be performed by the CREATE PURCHASE CONTRACTS module that originally captures contract data will also have to be performed by the CORRECT PURCHASE CONTRACTS module. This is particularly true for editing routines. Since that's the case, it should be efficient from a programming point of view to combine the functions into one module. On the other hand, once one of the programs is coded and tested, it's easy to make a copy of it and modify the copy to create the second program.

In contrast to an interactive system, machine efficiency may be important in a batch system. If, for example, an entire file is read sequentially five times to prepare five different reports in a batch subsystem, it's more efficient for the file to be read only once by a program that prepares all five reports simultaneously. So packaging five system modules into one certainly makes sense in a case like this.

To summarize, I prefer to see the modules of the original system structure chart reflected in the final program packaging unless there's a good reason for repackaging. In an interactive subsystem, the reasons for repackaging are usually related to user perceptions of (1) how the functions ought to be grouped and (2) ease of operation. In a batch subsystem, processing efficiencies are the primary consideration.

When to divide system modules   In some cases, it is desirable to divide a single system module into several subordinates, each of which will be implemented as a separate program. To illustrate, consider the system structure charts in figure 8-3. They present two ways in which the CREATE PURCHASE CONTRACTS module of the brokerage system could be implemented. In the first way, module BF1110 of the original system structure chart is implemented as one program. In the second way, this module is implemented as a menu program

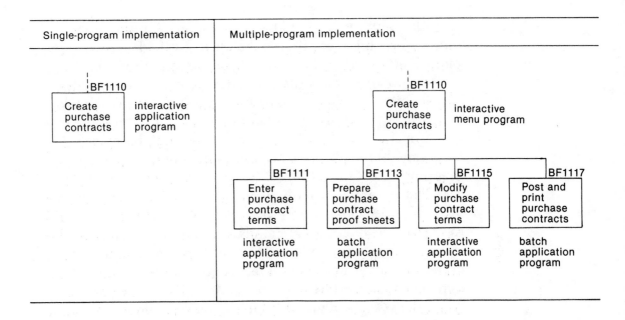

Figure 8-3   Two packaging alternatives for the CREATE PURCHASE
CONTRACTS module of the brokerage system

that controls the execution of four subordinate programs.
Both ways are acceptable implementations. It all depends on
the related factors.

In the first implementation in figure 8-3, one program
accepts contract data, adds new contracts to the purchase
contracts data store, prints an audit trail document that lists
the contracts just added to the data store, and prints the contracts themselves. In the second implementation, these same
functions are done, but they are done in separate program
steps. First, the user enters new contract data (BF1111), but
new contracts are *not* added to the contracts data store.
Second, the system prints a proof sheet of all contracts just
entered (BF1113) so the user can proof the entries against the
source documents. Third, if the user finds mistakes, he can
modify the original entries (BF1115). Fourth, when the user is
satisfied that all entries are correct, they're added to the contracts data store, the audit trail listing is printed, and the contracts are printed (BF1117). In this second implementation,
the user controls the execution of the four programs via a
menu (BF1110).

Although the second alternative requires more programming effort, there are times when it is justified. If, for example, it's critical that the contracts be correct before they're added to the contracts file, the extra proofing step may be worthwhile. We use a similar process when we enter book orders into our own system because we want them to be error free. On the other hand, when a system's database needs to be updated immediately, this alternative may be inappropriate.

This example shows how easy it is to divide a system module into two or more programs. You may want to do this to change the operational details of a module as illustrated by this example. You may want to do this to simplify the programming requirements of a large module so it is easier to implement the required functions. And, if your computer limits the size of programs, you may have to do this in order to implement a system module.

How to use procedures when packaging programs   In a menu-driven system, most of the selections invoke interactive programs. However, some selections may invoke *procedures* that control the execution of batch programs.

In figure 8-3, for example, module BF1117 could be packaged as two programs: one to update the purchase contract file and another to prepare the audit trail document and the new contracts. In this case, the update and the print programs need to be executed as part of the same procedure. If either program is executed and the other is skipped, there'll be a problem. Although it would be possible to execute each program by a menu selection and force the operator to choose one after the other in the right sequence, that's a waste of operator time, and it's error-prone to boot. Instead, the programs should be executed as part of a procedure.

An example of a module in the brokerage system that requires the use of a procedure is module BF1410 in part 5 of figure 6-8. Here, a listing of all payables on which the brokerage owes money is to be printed. But as I discovered when I developed the program specifications for this module, the payable items need to be sorted into the proper sequence to prepare this report. As a result, I had to implement this module with a procedure that first sorted the file, then executed the program to prepare the report. (My system didn't provide for sorts within application programs, though large systems usually do.)

When you name a procedure, you should assign a name that fits into the numbering scheme of the system structure chart. In the brokerage system, for example, I named the procedure I've just described BF1410. That's the same name the application program it invokes has, because our computer system allows a procedure and a program to have the same name. If that's not allowed on your system, you might try qualifying the procedure name. For example, you could name this procedure BF1410P where the suffix "P" indicates that the file is a procedure.

As a standard practice, I don't show system utility programs that are parts of procedures on the system structure chart. As a result, I didn't need to change the chart in figure 6-8 to reflect the fact that BF1410 will be a procedure that invokes more than one program. However, if two application programs are part of the same procedure, that should be indicated on the system structure chart.

How to design reports

Developing a report layout is a relatively straightforward task. As long as you follow consistent formatting and labelling practices and use common sense when you design an item of printed output, you really can't go wrong.

Consistent formatting makes work easier for the clerks who will distribute and file the reports your programs create. Consistent formatting implies that report titles appear in the same places on all reports and that those titles are clear. The people who use the reports will also appreciate consistent formatting and labelling.

The standard report format I used for the brokerage system is illustrated by the cash receipts register in figure 8-4. Each report prepared by the system has its title centered and the same identifying data elements in the same positions on the first lines of the report. No matter how you format your reports, every report should be labelled with a title, its preparation date, its preparation time (if appropriate), and

ABC BROKERAGE, INC.

C A S H   R E C E I P T S   R E G I S T E R

CUSTOMER	CHECK NUMBER	CHECK AMOUNT	CONTRACT	INVOICE	ORIGINAL BALANCE	PAYMENT APPLIED	DISCOUNT GRANTED	REMAINING BALANCE
WEST COAST MILLS	52234	32,701.01	00966-0	4321	32,701.01	32,701.01	.00	.00
VALLEY FARMS	922	1,200.00	00118-0	4270	2,284.17	1,200.00	.00	1,084.17
SMITH RANCHES	002144B	14,868.99	00302-0	4294	6,099.73	6,099.73	.00	.00
			00302-0	4301	3,533.22	3,533.22	.00	.00
			00302-0	4372	5,237.04	5,236.04	1.00	.00
		----------					--------	
		48,770.00					1.00	

Figure 8-4   A sample report: the brokerage system's cash receipts register

some unique identifying number. I recommend that you use a program's structure chart number to identify the reports it produces. For example, the cash receipts register in figure 8-4 is printed by program BF3410 in the system structure chart (figure 6-8, part 10). Because this report is labelled BF3410-1, it implies that program BF3410 prepares more than one output print item. The other is a bank deposit slip, and its number is BF3410-2. (The numbers of reports that are the only printed output of a program need not be qualified with -1.) If you follow this numbering scheme, you'll make your system more self-documenting.

In some instances, it is desirable to assign serial numbers to reports to insure that all copies of it are accounted for. For example, to provide internal controls, program BF3410 assigns a serial number to each copy of the cash receipts register it prepares. In figure 8-4, the register is number 527. Then, if the most recent copy in the register file is 525, it's immediately apparent that copy 526 is missing. The problem might be with clerical procedures or with the system. But the gap in the serial numbering sequence points out an error so steps can be taken to correct it.

Notice that all of the columns of the report in figure 8-4 are labelled. This is an obvious point, but all data elements on printed output should be identified. If the format of the report is non-columnar, you should use captions to identify data elements.

To design a report, use a report layout form (or print chart). For instance, figure 8-5 shows the report layout form for the cash receipts register shown in figure 8-4. You can see that the positions of all data elements are indicated on the form. Literal values (headings and titles) are written just as they will appear on the printed report. Variable data is specified by the editing characters of the language in which the program that prepares the report will be written (in this case, COBOL). Working from this report layout and the accompanying program overview for BF3410, the programmer assigned to code the program should have all the information necessary to do the job.

As a practical matter, when you've decided what the standard heading of a report should look like, you should code the required field definitions and make them available to

Document name Cash Receipts Register   Date 5-20-83

Program name BF3410   Designer SLE

Record Name

1	DATE:	XX/XX/XX	PAGE: 22229
2	TIME:	XX:XX.XX	BF3410-11
3	SERIAL NO:	22229	
4	CASH RECEIPTS REGISTER		
5	ABC BROKERAGE, INC.		

CUSTOMER	CHECK AMOUNT	CONTRACT	INVOICE	ORIGINAL BALANCE	PAYMENT APPLIED	DISCOUNT GRANTED	REMAINING BALANCE
CHECK NUMBER							
XXXXXXXXXXXXXX	2,222,222.99	99999-9	99999	22,222.99-	22,222.99-	22,222.99-	22,222.99-
XXXXXXXXXXXXXX	2,222,2222.99	99999-9	99999	22,222.99-	22,222.99-	22,222.99-	22,222.99-
		99999-9	99999	22,222.99-	22,222.99-	22,222.99-	22,222.99-
	2,222,2222.99-						22,222.99-

Figure 8-5 The print chart for the cash receipts register

all programmers. Then, they can be copied into all report-preparation programs. This has the double benefit of forcing uniformity and cutting down on coding time.

How to design screens for interactive programs

Most new systems under development today are interactive. And many batch systems are being converted to interactive systems. As a result, more and more users operate interactive workstations, or terminals. To improve their productivity, workstation screens should be designed so they're as easy as possible to work with.

If all screens in a system are formatted in a standard way, two benefits will result. First, the users will find that the interactive programs are easier to use. Second, it will be easier to create and maintain the programs that display and accept the screens. As a result, if you haven't already done so, you should develop a set of standards for screen design that emphasize consistency and ease of operation. Some suggestions for these standards follow.

Use a screen layout form or its equivalent   Much of the preliminary work of screen design can be done on scratch paper. However, when you create a final screen design for a program, it should be on a screen layout form or its equivalent. The screen layout form looks much like a report layout form, but its rows and columns correspond to the dimensions of the screen of the workstation on which the application will be run. For example, figure 8-6 illustrates a typical screen design on a screen layout form. Notice that this form has 80 columns and 24 lines. Those are the standard dimensions of a workstation screen. If you are creating screen designs for a workstation with a larger or a smaller screen, you should adjust the size of the screen layout form accordingly.

If your system has screen-formatting utilities, you may not need a manual form like the one in figure 8-6. These utilities make it easier for a designer or programmer to prepare screen layouts. So if they're available, you should use them. At the least, they make it easier to plan what a screen should look like because you do the planning on a screen, not

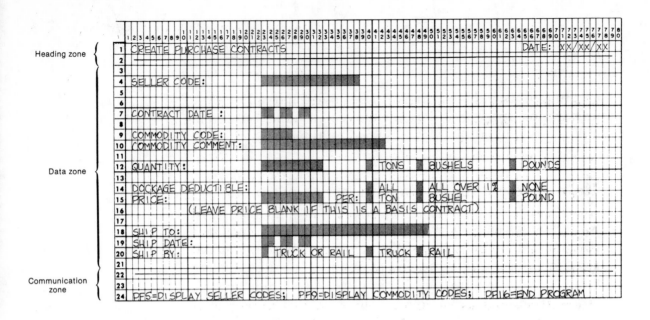

Figure 8-6 A screen layout for one screen of the CREATE PURCHASE CONTRACTS program in the brokerage system

on a piece of paper. In addition, some utilities let you save a screen image so you can recall it and modify it later on. For documentation of the screen, some utilities print the screen along with row and column numbers, thus eliminating the need for a manual form.

Divide the screen into zones   In general, you should divide a screen into the three zones illustrated in figure 8-6. The first, the *heading zone*, includes the name of the program that displays the screen (in this case, CREATE PURCHASE CONTRACTS). The heading zone may also contain additional information, such as a date or a screen number if the program displays more than one screen.

The second, the *data zone*, contains all of the data relevant to the application. In figure 8-6, that includes the entry fields and captions necessary for the contract-creation process. In an inquiry program, the data zone might not contain any entry fields. When you design a screen, most of your time will be spent on the data zone.

The third, the *communication zone*, is used to display error messages and operator instructions. I dedicate the last two lines of the screen for this zone (typically, lines 23 and 24).

Make the screen self-documenting   This means you should do all you can to make your interactive programs as self-contained as possible. Don't force the operator to refer to manuals or written instructions to figure out how to work a program. Instead, all user instructions and options should be clearly specified on the screen itself or on an easily accessible reference screen. All elements displayed on the screen should be clearly labelled. And all error messages should be clear and unambiguous so they guide the operator through an entry. Soon, I'll have more to say about this.

Align data elements vertically   One of the basic rules of effective screen design is to align fields vertically on the screen. To illustrate, consider the following fields from figure 8-6:

```
SELLER CODE:            XXXXXXXXXXXXXXXX

CONTRACT DATE:          99 99 99

COMMODITY CODE:         XXXXX
COMMODITY COMMENT:      XXXXXXXXXXXXXXXXXXXX

QUANTITY:               999999999
```

Now consider how much more difficult they are to read when they are not aligned vertically:

```
SELLER CODE: XXXXXXXXXXXXXXXX

CONTRACT DATE: 99 99 99

COMMODITY CODE: XXXXX
COMMODITY COMMENT:      XXXXXXXXXXXXXXXXXXXX

QUANTITY: 999999999
```

Not only does aligning fields improve readability, but it also makes the screen easier to use. In this example, for instance, an operator can move the cursor from the COMMODITY CODE field to the COMMODITY COMMENT field more easily when the fields are properly aligned.

Group data elements logically   You should also improve the readability of your screens by grouping logically related items

and separating them from other groups. You can separate
groups with blank lines or by positioning them in different
vertical zones of the screen. In figure 8-6, the items
COMMODITY CODE and COMMODITY COMMENT are
logically related, so I grouped them. Similarly, the dockage
and pricing fields are related, and the shipping fields are
related, so I grouped them too. The remaining data elements
aren't related, so I separated them from one another with
blank lines.

   In contrast, see how difficult the screen would be to use
if these items were neither vertically aligned, nor grouped
logically:

```
SELLER CODE: XXXXXXXXXXXXXXX
CONTRACT DATE: 99 99 99
COMMODITY CODE: XXXXX
QUANTITY: 999999999
COMMODITY COMMENT:      XXXXXXXXXXXXXXXXXXXX
```

Don't force logical groupings where they don't exist, though.
If your screen consists of 15 unrelated items, leave it at that.
Or, more likely, if you have three or four clear groupings and
you're left with one or two data elements that don't belong in
any of the groups, leave them by themselves.

Don't crowd a screen unnecessarily   Many applications
require more fields than one screen can readily accommodate.
If so, don't crowd one screen. Use multiple screens. When you
split data over several screens, never split up a logical grouping of data. Use the groupings as a guide to help you determine where to split the screens.

Use highlighting consistently   Most workstations provide
some way to highlight areas on a screen. Highlighting can be
used effectively to draw the operator's attention to entry
areas, displayed data, and messages. Just be consistent in your
use of highlighting. Follow a standard that defines what types
of fields to display with normal intensity and what types of
fields to display with high intensity.

Indicate errors in a consistent and useful way   An error
message should always contain two elements: (1) an iden-

tification of the field in error, and (2) a brief yet complete explanation of what's wrong. For example, a message like:

INVALID FIELD

is an unacceptable message if it doesn't tell the operator which field is invalid or if it's not obvious why the field is invalid. A message like:

SELLER CODE NOT FOUND IN SELLER FILE

is better. It tells which field is in error and what's wrong with it. On the other hand, use common sense. Don't go overboard about creating a unique error message for every possible situation when one more general message will do the job.

If, for example, a date field is invalid, there's no need to indicate whether the month, day, or year portion of the field is bad. Just say the date's invalid, and give the format for a correct date, like this:

INVALID DATE MUST BE IN THE FORM MM/DD/YY

Since the user should be able to understand this message, the message is satisfactory.

Be consistent in your indication of errors. Display error messages in the same place on the screen all the time. I recommend the second to the last line on a screen (that's usually line 23). Identify the field in error in the message and, if possible, highlight that field in the data zone of the screen to make it obvious to the operator. If several fields are in error, display a message that explains what's wrong with the first one in error on the screen, but highlight all of the incorrect fields. That way, the operator has the chance to correct more than just the error to which the message refers.

Provide consistent and complete operator instructions whenever possible   In order to make programs as easy as possible to use, every screen should provide a clear indication of all processing options open to the operator. And those options should be consistent from one program to another. At the least, the screen should indicate the function of each of the workstation's attention keys such as the clear key and the program function (PF) keys. All workstations use an enter (or return) key, and most support some other combination of attention keys. So, at this point, you should standardize the functions for those keys.

The screen in figure 8-6 indicates that three attention keys are in operation. The PF5 key will cause a reference screen of valid seller codes to be displayed. The PF9 key will cause a reference screen of valid commodity codes to be displayed. And the PF16 key will cause the program to terminate. To make the instructions more concise, you shouldn't display the function of the enter key because it should be obvious.

Consistency from one program to another is important here too. As a system standard, all of our programs use the PF16 key to end a program or to escape from an entry cycle. Similarly, when we ask an operator to validate displayed data, PF1 indicates that the data is acceptable, and PF4 indicates that it should be modified. Because all of the programs at our installation use these conventions, our operators don't have to learn minor variations from one program to another, so training is more efficient. Also, our screen-handling programs are easier to develop and maintain.

Provide reference screens whenever appropriate   If you're concerned about operator efficiency, you should try to make your interactive programs as self-documenting as possible. In other words, the operator should not have to refer to additional reference material in order to use an interactive program. One of the best ways to achieve this goal is to provide reference screens whenever needed. A *reference screen* explains the options that are available to an operator or displays an array of valid codes when there are too many to display on the entry screen itself. The reference screen is displayed only when the operator needs to see it, so it doesn't interrupt the normal flow of program operation.

In figure 8-6, for example, two fields require the entry of a code: SELLER CODE and COMMODITY CODE. If the operator doesn't know what the valid codes are, he can press a function key and review a list of all valid seller or commodity codes (PF5 and PF9 respectively). For instance, figure 8-7 shows the reference screen for commodity codes displayed if the user selects PF9.

```
CREATE PURCHASE CONTRACTS                              DATE: 06/24/83

   CODE     DESCRIPTION              CODE       DESCRIPTION
   515AS    ALFALFA SEED WL 515      CB         CULL BEANS
   572AS    ALFALFA SEED, PIONEER 572 CC        CORN CRACKS
   581AS    ALFALFA SEED 581         CC-1       CORN CRACKS
   819AS    ALFALFA SEED 819         CC-2       CORN CRACKS
   819BL    ALFALFA SEED 819 BLENDED CC-BP      CORN CRACKS
   AH       ALMOND HULLS             CCM72      BARLEY SEED - CERT CM72
   AH-S     ALMOND HULL & SHELL      CDF        CULLED DRIED FIGS
   ALF H    ALFALFA HAY              CDP        CULLED DRIED PRUNES
   APG      APPLE, PEAR & GRAPE POMACE CM72     BARLEY SEED CM 72
   AS       ALMOND SHELL             COS        CAYOUSE OAT SEED
   ASS      ALFALFA SEED SCREENINGS  CROS       OAT SEED CALIFORNIA RED
   B        BARLEY                   CS318      CORN SEED
   BP       BEET PULP                CSF13      CORN SEED
   BPP      BEET PULP PELLETS        CSH        COTTONSEED HULLS
   C        CORN                     CSM        COTTONSEED MEAL
   CAR      CARROTS                  CUFF       ALFALFA SEED CUFF 101


 ENTER=DISPLAY MORE COMMODITY CODES; PF16=RETURN TO ENTRY SCREEN
```

Figure 8-7   A reference screen for the CREATE PURCHASE
              CONTRACTS program

Of course, you must use your own judgement when
creating reference screens. Depending on the application, a
reference screen like the one in figure 8-7 may or may not be
helpful. If, for example, the brokerage deals in 5,000 different
commodities and the commodity code is always clearly indicated on the contract specifications, a reference screen is
both impractical and unnecessary. On the other hand, if the
firm deals in 100 different commodities and they change frequently, then the reference screen can save the operator from
looking the codes up in a reference book, perhaps providing a
significant improvement in operator efficiency.

In reality, the brokerage firm deals in about 75 commodities, and they don't change. As a result, the reference
screen in figure 8-7 is rarely used. It's only accessed when a
new operator is learning the codes or when an experienced
operator must create a contract for an infrequently purchased
commodity. This reference screen, then, probably isn't
necessary, but it is convenient.

In summary, you must use your own judgement when
deciding whether to provide a reference screen. But be sure to
consider all possible reference screens, even if you eventually
decide not to include any of them.

Avoid using the special features of workstations   Some
workstations have special features, such as underlining, color,
audio alarm, light pen, blinking, and so on. Although the
temptation to use these features is great, avoid it. They don't
really contribute much to the effectiveness of your interactive
programs, and they can complicate program development
considerably. If you do use some of these features, be sure to
establish standards as to how and when they will be used, so
they'll be used consistently throughout your installation.

**Special guidelines for screens with a dedicated source
document**   A *dedicated source document* is an input source
document that is always used with a particular data-entry
program. For example, an insurance application form is a
dedicated source document for an insurance application entry
program. When an operator is using a dedicated source document with a data-entry program, he usually doesn't look at
the screen except to correct errors and verify data. When he
does look at the screen, he should find that it closely corresponds to the source document. So when you design a
screen that has a dedicated source document, follow the
source document as closely as you can.

   When you pattern the screen after a source document,
you sometimes have to compromise the vertical alignment in
order to maintain a close relationship with the source document. If you're designing the source document at the same
time that you're designing the screen, of course, you can
create an effective screen design and make the source document match it.

   If you pattern your screen closely after the source
document, you can make your captions relatively brief. For
example, a caption like

```
COMM:
```

is acceptable if, by the context of the source document, it's
clear that this field represents commodity. In fact, since most
source documents use a small type size for captions, it's almost
impossible to follow the source document if you don't shorten
the captions.

For screens with a dedicated source document, coded entry is acceptable as long as the code value is indicated clearly on the source document. Also, since most forms indicate the valid values for the code, there is usually no reason to show the valid values on the entry screen or on a related reference screen. That would just take up extra space on the screen, and the operator would never look at it anyway.

Special guidelines for screens without a dedicated source document   When an entry screen doesn't have a dedicated source document, you should try to pattern the screen after a related output or source document. For example, the contract-entry screen in figure 8-6 is patterned after the contract produced by the order-entry program. The data items on the screen appear in the same sequence as the data items on the printed contract. If there is no output document produced from the input data, pattern the screen after a typical source document.

When the screen doesn't have a dedicated source document, you must be sure that all of the captions on the screen clearly identify their data fields. So you shouldn't use abbreviations that aren't obvious. Abbreviations like NO for NUMBER or CUST for CUSTOMER are usually clear, but others may not be. In figure 8-6, for instance, I spelled out COMMODITY because I felt COMM could easily be confused with other entry fields.

For fields displayed by the program, expand the meaning of coded data. For example, an inquiry program should display

COMMODITY: YOCORO ROJO WHEAT SEED

rather than

COMMODITY: YRWS

The result is a screen that is much more meaningful to the operator.

For entry fields, don't use coded values if you can avoid them. The best way to eliminate coded values is to list all possible values with a one-byte entry field next to each. In figure 8-6, the operator indicates the quantity and pricing units and how the purchased commodity should be shipped by placing a non-blank character next to the desired options.

If this isn't practical (perhaps too many codes are possible), the operator will have to enter the code. But you should still try to provide an explanation of all the valid codes on the screen, like this:

```
SHIP BY: T (E=EITHER TRUCK OR RAIL,
            T=TRUCK, R=RAIL)
```

If there are more values than can be shown easily on the same screen, you should consider providing a reference screen that shows all of the valid codes. Then, the operator can display the reference screen when necessary.

How to plan
the screen flow of an interactive program

The design of a traditional batch program is fairly straightforward and is generally left to the programmer. In contrast, the design of processing flow in an interactive program is more critical. As a result, you should develop standards for screen flow that should be followed in all interactive programs at your installation. That means general decisions need to be made by the system designer or design team.

Of course, all programs, both batch and interactive, should edit all input data thoroughly. But even the most extensive editing routines leave room for entry errors. So the responsibility for entry accuracy must be given to the entry operators. And your entry programs should help them be as responsible as possible.

As an ideal, interactive programs should do all they can to make sure that operators visually verify data before it is added to a transaction file or used to update a master file. In actual practice, however, operators often release entries for processing without visually reviewing them. They're too anxious to complete the workload for the day.

As a standard for interactive program flow, I'm going to recommend a process that requires a positive response by the operator after visually verifying each screen. In our installa-

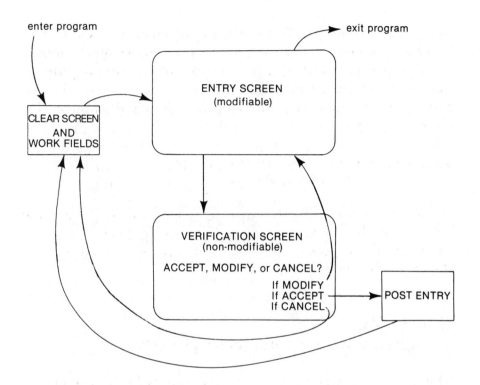

enter program

exit program

ENTRY SCREEN
(modifiable)

CLEAR SCREEN
AND
WORK FIELDS

VERIFICATION SCREEN
(non-modifiable)

ACCEPT, MODIFY, or CANCEL?

If MODIFY
If ACCEPT
If CANCEL

POST ENTRY

Figure 8-8   A suggested screen flow for a single entry-screen program

tion and at the brokerage firm's, entry accuracy is more important than entry cost. But feel free to simplify this standard based on your application, your clerical environment, your computer system, or your budget. For example, the computer system on which you'll be implementing your applications may not lend itself to programs that work like ours. Also, a program that does the thorough verification that I'm going to suggest is more costly to develop than one that doesn't, so that may be the bottom line for you. Nevertheless, it's important that you adopt a standard for screen flow, so I hope you'll use my suggestions as a starting point.

Figure 8-8 illustrates the flow I recommend for a program that uses a single entry screen. To start, the program should edit all the data entered by the operator as thoroughly as possible. When the entered data is edited without errors, the program should display a verification screen that redisplays the entered data. When the data is redisplayed, coded data is also expanded. For example, the verification screen will expand a buyer code into buyer name and address so the operator can verify that he has identified the right buyer. Usually, the verification screen will have approximately the same format as the entry screen, but the operator will not be able to modify any of its fields.

The purpose of the verification screen is to give the operator a chance to check over the entries a second time to make sure that they are correct. If they are, the operator can accept the data, using one of the attention keys. Once accepted, the data is posted to related files, the program's work fields are reset for the next transaction, and the entry screen is cleared. But if the entries aren't correct, the operator can modify them or cancel the transaction altogether. If modify is the selection, the program returns directly to the entry screen so the operator can change the incorrect entries. If cancel is the selection, the program clears the work fields and the entry screen, then returns to the entry screen.

Figure 8-9 illustrates this screen flow as implemented in the CREATE PURCHASE CONTRACTS program. I chose this example because it illustrates the use of a verification screen, but it doesn't exactly match the flow I just described for a single entry-screen program, nor the one I'm about to describe for a multiple entry-screen program.

In this example, the normal program flow will be like that in figure 8-8. To start, the screen in part 1 of figure 8-9 is displayed and filled in by the user. This is the only entry screen for a fixed price contract. For this type of contract, a specific price is set at the time the contract is created, and almost all contracts are fixed price contracts.

However, a second type of contract (called a basis contract) is occasionally written that is not for a fixed price. Instead, the price that is paid for the contracted commodity is determined by adding or subtracting a specified amount to the price of a specified commodity at a specified date at a specified market location. So all of that specification is required to create a basis contract.

To provide for a basis contract, then, the CREATE PURCHASE CONTRACTS program uses a second entry screen as shown in part 2 of figure 8-9. Since most contracts are written with fixed prices, though, this second screen only appears when needed. It's displayed when the user doesn't enter a fixed price on the program's primary entry screen.

```
  1 2 3 4 5 6 7 8 9 0 1 2 3 4 5 6 7 8 9 0 1 2 3 4 5 6 7 8 9 0 1 2 3 4 5 6 7 8 9 0 ...
 1 │CREATE PURCHASE CONTRACTS                              DATE: XX/XX/XX
 2 │
 3 │
 4 │SELLER CODE:         ███████████
 5 │
 6 │
 7 │CONTRACT DATE:       ██ ██ ██
 8 │
 9 │COMMODITY CODE:      ████
10 │COMMODITY COMMENT:   █████████████████
11 │
12 │QUANTITY:            ████████      ▓ TONS   ▓ BUSHELS      ▓ POUNDS
13 │
14 │DOCKAGE DEDUCTIBLE:                ▓ ALL    ▓ ALL OVER 1%  ▓ NONE
15 │PRICE:               █████ PER:    ▓ TON    ▓ BUSHEL       ▓ POUND
16 │      (LEAVE PRICE BLANK IF THIS IS A BASIS CONTRACT)
17 │
18 │SHIP TO:             █████████████████
19 │SHIP DATE:           ██ ██ ██
20 │SHIP BY:             ▓ TRUCK OR RAIL  ▓ TRUCK  ▓ RAIL
21 │
22 │
23 │
24 │PF5=DISPLAY SELLER CODES;  PF9=DISPLAY COMMODITY CODES;  PF16=END PROGRAM
```

Screen layout for the PURCHASE CONTRACT ENTRY SCREEN

```
┌─────────────────────────────────────────────────────────────────────┐
│                                                                       │
│   CREATE PURCHASE CONTRACTS                        DATE: 06/24/83      │
│   ─────────────────────────────────────────────────────────────────  │
│                                                                       │
│   SELLER CODE:         XYZ_____                                    │
│                                                                       │
│   CONTRACT DATE:       6  24 83                                        │
│                                                                       │
│   COMMODITY CODE:      W___                                            │
│   COMMODITY COMMENT:   12% PROTEIN OR MORE                             │
│                                                                       │
│   QUANTITY:            500_____    X TONS    _ BUSHELS    _ POUNDS   │
│                                                                       │
│   DOCKAGE DEDUCTIBLE:                X ALL     _ ALL OVER 1%  NONE     │
│   PRICE:               _____ PER: _ TON    _ BUSHEL     _ POUND    │
│        (LEAVE PRICE BLANK IF THIS IS A BASIS CONTRACT)                 │
│                                                                       │
│   SHIP TO:             STOCKTON_____                                   │
│   SHIP DATE:           12 15 83                                        │
│   SHIP BY:             _ TRUCK OR RAIL X TRUCK    _ RAIL               │
│   ─────────────────────────────────────────────────────────────────  │
│                                                                       │
│   PF5=DISPLAY SELLER CODES; PF5=DISPLAY COMMODITY CODES; PF16=END PROGRAM │
└─────────────────────────────────────────────────────────────────────┘
```

PURCHASE CONTRACT ENTRY SCREEN as displayed with operator-entered data

Figure 8-9   An example of screen flow in the CREATE PURCHASE
            CONTRACTS program (part 1 of 3)

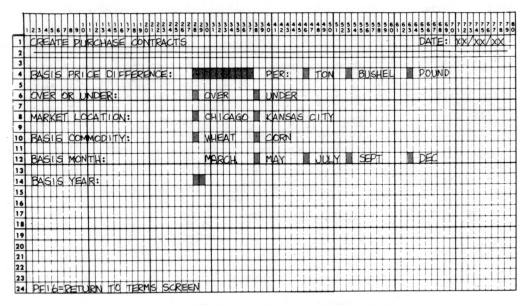

Screen layout for the PURCHASE CONTRACT BASIS PRICING ENTRY SCREEN

```
CREATE PURCHASE CONTRACTS                        DATE: 06/24/83
_____

    BASIS PRICE DIFFERENCE:     1.5          PER: X TON  _ BUSHEL  _ POUND

    OVER OR UNDER:              X OVER         _ UNDER

    MARKET LOCATION:           _ CHICAGO     X KANSAS CITY

    BASIS COMMODITY:           X WHEAT        _ CORN

    BASIS MONTH:               _ MARCH       _ MAY _ JULY  _ SEPT   X DEC

    BASIS YEAR:                83


    _____

    PF16=RETURN TO TERMS SCREEN
```

PURCHASE CONTRACT BASIS PRICING ENTRY SCREEN as displayed with
operator-entered data

Figure 8-9   An example of screen flow in the CREATE PURCHASE
CONTRACTS program (part 2 of 3)

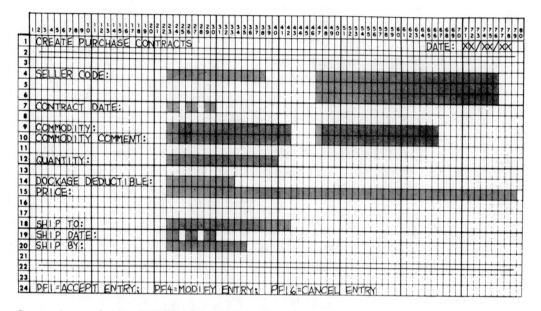

Screen layout for the PURCHASE CONTRACT VERIFICATION SCREEN

```
  CREATE PURCHASE CONTRACTS                          DATE: 06/24/83
  _____

     SELLER CODE:           XYZ                   XYZ FARMS, INC.
                                                  5472 AVENUE 9
                                                  FRESNO, CA 93745


     CONTRACT DATE:         06 24 83

     COMMODITY:             WHEAT                  HARD RED WINTER
     COMMODITY COMMENT:     12% PROTEIN OR MORE    U.S. GRADE #2

     QUANTITY:              500 TONS

     DOCKAGE DEDUCTIBLE:    ALL
     PRICE:                 $1.50 PER TON OVER KANSAS CITY WHEAT DECEMBER 83 OPTION

     SHIP TO:               STOCKTON
     SHIP DATE:             12 15 83
     SHIP BY:               TRUCK

  _____

   PF1=ACCEPT ENTRY;  PF4=MODIFY ENTRY;  PF16=CANCEL ENTRY
```

PURCHASE CONTRACT VERIFICATION SCREEN with operator-entered data
and program-supplied supplementary data

Figure 8-9   An example of screen flow in the CREATE PURCHASE
CONTRACTS program (part 3 of 3)

From the perspective of the user, this is a one-screen program that in special circumstances requires additional entry on a supplementary screen. The result is that I designed the screen flow for this program along the lines of the one entry-screen model in figure 8-8. I only coded one verification screen as shown in part 3 of figure 8-9, but the format of the price information varies based on whether or not the new contract has basis pricing. If the operator decides to modify the entries, he can change from one type of contract to the other by specifying or not specifying a fixed price.

If a program typically requires more than one entry screen, the screen flow is complicated somewhat. Each entry screen should have its own verification screen, just as in the single entry-screen model. In addition, a final verification screen should be used to advise the operator that the entry has been completed and is ready for posting. Figure 8-10 illustrates a model screen flow for a three entry-screen program. For all but the last set of screens, if the entered data is accepted, the next entry screen in the sequence is displayed. However, the operator may cancel the transaction at any verification screen in the progression.

Of course, most programs require only one or two entry screens. If you find yourself planning a program that requires many screens, you should probably review what you're trying to do and make sure it's a reasonable function. That's not to say, though, that a program shouldn't have many screens. In the brokerage system, the program that corresponds to BF2100 in the system structure chart in figure 6-8 (ENTER DELIVERIES) uses 28 separate screens, and it works logically and effectively. But, it's the exception, not the rule.

As I said, these are just models for screen flow within interactive programs. In actual practice, these models may be too expensive to implement in your shop. Or they may require more internal storage than your computer can provide. However, for critical applications in which the data that is entered must be correct, I suggest that you use models like this. And I suggest that you enforce visual verification by your operators.

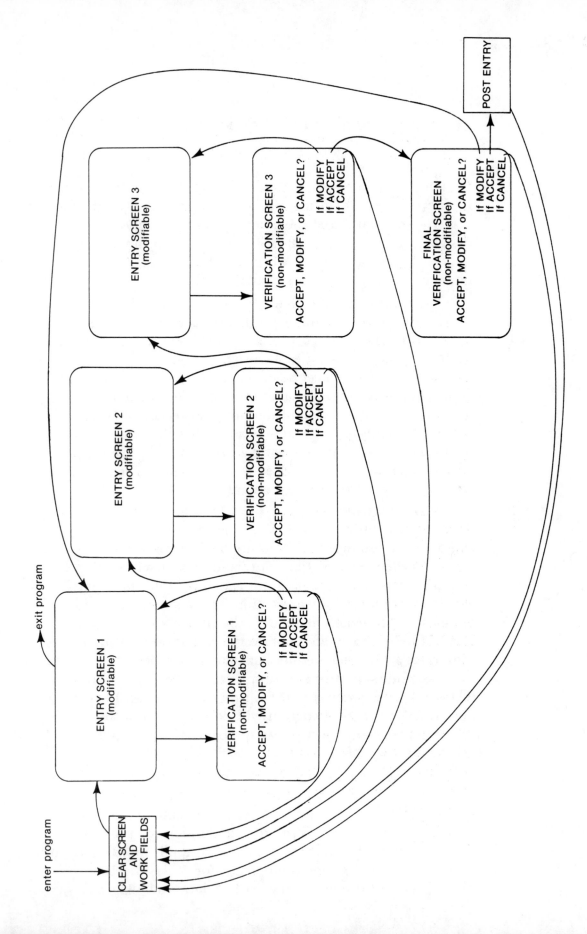

Figure 8-10 A suggested screen flow for a multiple entry-screen program

How to create program overviews

Creating complete specifications for each program in a system is an extensive, exhausting job. This is aggravated by the fact that you're usually past the creative work of analysis and design by this time. So you're left to deal with the details of how each program should work. Need I say that this is the phase of development work that I like the least.

With this in mind, what I want to present now is a way to create effective program specifications in the shortest possible time. In our experience, the most efficient program documentation, both for designer and programmer, is what we call the *program overview*. It's a listing of the input and output requirements of a program, combined with a brief, but complete, description of the program requirements. If that means using pseudocode in the overview, we use pseudocode. If that means using a decision table, we use a decision table. And so on. Even an occasional flowchart or section of narrative text is acceptable.

Figure 8-11 illustrates a typical program overview as developed for program BF1110 of the brokerage system. Although we use the form in figure 8-11 for program overviews, I have seen variations of this form that will work just as well. As you can see, the form is divided into three parts. The top section is for identification. It gives the name and number of the related module on the system structure chart. Then, it tells who created the overview and on what date. Finally, it gives the page number for the overview.

The middle part of the form lists and describes all of the files (or database views) that the program requires. In addition, it tells what the program will do with each of those files: use them only for input, use them only for output, or use them for update.

The last section of the form is for processing specifications. It's this section, of course, that takes the most time to complete. Here, you need to give all the detailed information that a programmer needs in order to develop the program. How you do that is up to you. For most applications, a sort of non-rigorous pseudocode is efficient. For others, decision tables, partial flowcharts, and partial data flow diagrams

Input/output specifications

File	Description	Use
PURCHCON	Purchase contract master file	Update
SELLERS	Seller reference file	Input
COMMODTY	Commodity reference file	Input
PCCNTRL	Purchase contract control file	Update
PCONTRCT	Print file: purchase contract copies	Output
PCREG	Print file: purchase contract register	Output

Process specifications

Get the system date using the subprogram SYSDATE

Get the system time using the subprogram SYSTIME

Do until operator indicates end of program by pressing PF16 from PURCHASE CONTRACT ENTRY SCREEN (in step 1):

1. Accept PURCHASE CONTRACT ENTRY SCREEN until no errors are detected in operator-entered fields or operator indicates end of program. Field editing rules are below. If the operator indicates end of program, terminate program immediately.

2. If the operator enters no price data on PURCHASE CONTRACT ENTRY SCREEN, accept PURCHASE CONTRACT BASIS PRICING SCREEN until no errors are detected in operator-entered fields. Field editing rules are below.

3. Format the PURCHASE CONTRACT VERIFICATION SCREEN:

   —Display the name and address of the seller specified by the operator.

   —Display the full description of the commodity specified by the operator (name, variety, and quality).

   —Format lines where the operator had several options from which to choose so they display only the one selected.

   —If the entry is for a basis contract, combine the pricing parameters selected from the PURCHASE CONTRACT BASIS PRICING SCREEN and format them as a sentence.

Figure 8-11   A program overview for the CREATE PURCHASE CONTRACTS program (part 1 of 3)

Process specifications

4. Accept the PURCHASE CONTRACT VERIFICATION SCREEN.

   If the operator selects PF16, bypass step 5 and proceed with step 6.

   If the operator selects PF4, do neither step 5 nor step 6, but restart the entry cycle immediately.

   If the operator selects PF1, proceed with step 5.

5. Post the contract:

   —Read control file to get next purchase contract number

   —Update control file to roll numbers forward

   —Add new record to purchase contract file

   —Format and print a contract copy (per print chart)

   —Format and print a line on the purchase contract register (per print chart)

6. Clear all screen and work areas.

Editing rules for PURCHASE CONTRACT ENTRY SCREEN:

Seller code:	must be a valid code in the file SELLERS
Contract date:	must be a logically valid date in the format MM DD YY
Commodity code:	must be a valid code in the file COMMODTY
Commodity comment:	no restrictions
Quantity:	must be a positive, non-zero numeric value no greater than 999,999,999
Quantity units:	the user must select one of: tons, bushels, or pounds
Dockage deductible:	the user must select one of: all, all over 1%, or none
Price and units:	the user must leave the price field and all three price unit selections blank or must enter a positive, non-zero numeric value not greater than 999,999,999 for price and must select for units one of: ton, bushel, or pound
Ship to:	the user must enter a non-blank value
Ship date:	the user must enter a valid date in format MM DD YY that is on or after the contract date
Ship via:	the user must select one of: truck or rail, truck, or rail

Figure 8-11 A program overview for the **CREATE PURCHASE CONTRACTS** program (part 2 of 3)

Process specifications

Editing rules for PURCHASE CONTRACT BASIS PRICING ENTRY SCREEN:

Basis price difference and units:	the user must enter a positive non-zero numeric: value not greater than 999,999,999 for price and must select for units one of: ton, bushel, or pound
Over or under:	the user must select either over or under
Market location:	the user must select either Kansas City or Chicago
Basis commodity:	the user must select either wheat or corn
Basis month:	the user must select one among: March, May, July, September, or December
	must be numeric; when combined with basis month, must not be before the current month and year

Figure 8-11 A program overview for the CREATE PURCHASE CONTRACTS program (part 3 of 3)

may be useful. Often, you'll combine two or more of these methods in a program overview. As you can see by the three parts of figure 8-11, you can take as many pages as you need. Also, you can see that we use two different overview forms: one for the first page, one for subsequent pages.

If you have access to a word processing system, you can create your program overviews with it. Since much of the work of creating program overviews is repetitive, word processing can improve your productivity dramatically. In our shop, we don't actually use preprinted overview forms anymore. Instead, the forms are part of our word processing documents, so the rules for the form print along with the entries.

If you read the overview in figure 8-11, you should be able to understand it quite easily. If you read through the process specifications, you'll see that the first tasks the program must do are to get the system date and time using subprograms. Then, the program enters a cycle for creating contracts like the one in figure 8-9. This will continue until the operator signals that the program should be terminated (in our standards, by pressing the PF16 key). As I describe what happens in the entry cycle, I refer to the screen descriptions in figure 8-9 by name. When the specification is turned over to the programming staff for implementation, it will include the overview plus any report or screen layouts.

As much as possible, you should avoid detailed descriptions of how the program's processing should function when you create a program overview. This is the programmer's responsibility. However, it is essential that editing rules be spelled out. In figure 8-11, you can see that I've provided complete editing rules for all fields to be entered by the user. And if a program is to prepare reports or display summaries, the overview should completely define any calculation requirements for those output items.

You'll notice that I didn't obey any rigid rules for pseudocode in the process specification of the overview. My goal is simply to create as rapidly as possible the best specification I can create. It's not to follow a bureaucratic system of documentation for its own sake.

When you develop an overview, you may find that you need to add data stores that are internal to the process on which you're working. For example, in figure 8-11, I needed to add an internal control data store for BF1110 that maintains the next contract number. Because that data store is completely internal to the process, I didn't need to worry about its interfaces with the rest of the system.

Discussion

In our experience, the most efficient and effective way to create program specifications is to use the program overview as the basis of the specification. When combined with layouts for screens and reports and documentation for required files and subprograms, the overview is invaluable. It coordinates the other items of documentation and clears up specific questions about program functions. It tells the programmer what the program must do, but it lets the programmer decide how to do it.

Terminology

program packaging
procedure
heading zone of screen
data zone of screen
communication zone of screen
reference screen
dedicated source document
program overview

Objectives

1. Given system documentation, identify useful subprograms.

2. Given a system structure chart and related information, decide how its modules should be packaged into programs in the final system.

3. Given printed output requirements, design an effective report.

4. Given entry and display requirements, design an effective screen.

5. Describe a model for interactive screen flow that requires a response to a verification screen.

6. Given program requirements, create an effective program overview.

Part

Implementation

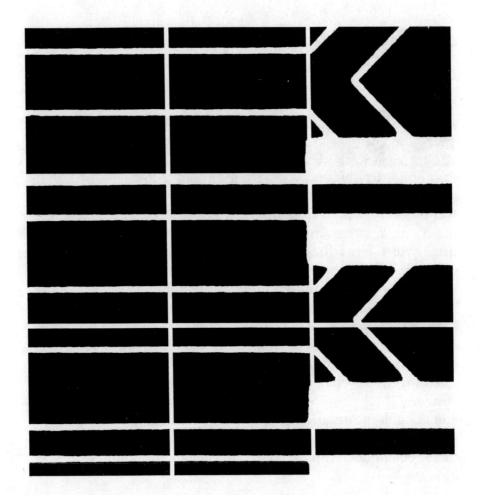

Chapter 9

How to develop programs
using structured programming techniques

I hope you develop all of your programs using structured programming techniques. If you do, you can consider skipping this chapter. On the other hand, structured programming means different things to different people. So you may be interested in the many ways that we apply structured programming techniques to program development.

If you're not familiar with structured programming techniques, it's high time that you get with it. These techniques make programming more enjoyable, improve the quality of the resulting programs, and increase programming productivity. So the sooner you start using these techniques, the better off your programming efforts will be.

In this chapter, I will first present the theory of structured programming. Then, I will show you how to apply this theory as you design, code, test, and document your programs. If you are already using structured programming in some form, this chapter should give you some new ideas for applying it. If you aren't, this chapter probably won't give you enough information to get you started using structured programming techniques. But it should motivate you to use these techniques. Then, you can master them by doing independent study.

The theory of structured programming

The basic theory of *structured programming* is that any program can be written using three logical structures: sequence, selection, and iteration. These structures, illustrated in figure 9-1, have only one entry point and one exit point.

The first, the *sequence structure*, is simply a set of imperative statements executed in sequence, one after another. The exit point is after the last function in the sequence. A sequence structure may consist of a single function or of many.

The second, the *selection structure*, is a choice between two, and only two, functions based on a condition. This structure is often referred to as the IF-THEN-ELSE structure, and most programming languages have code that approximates it. One of the functions may be null. In other words, if the condition is not met, the flow of control may pass directly to the structure's exit point with no intervening statements or structures.

The third, the *iteration structure*, often called the DO-WHILE structure, provides for doing a function as long as a condition is true. When the condition is no longer true, the program continues with the next structure. Logically related are the DO-UNTIL and the COBOL PERFORM-UNTIL functions.

Again, let me stress that all of the structures in figure 9-1 have only one entry point and one exit point. As a result, a program made up of these structures will have only one entry point and one exit point. Thus the program will be executed in a controlled manner from the first statement to the last. These characteristics make up a *proper program*.

To create a proper program, any of the three structures can be substituted for a function box in any of the other structures. The result will still be a proper program. Conversely, two or more of the basic structures in sequence can be treated as a single function box. The result? Structures of great complexity can be created with the assurance that they will have only one entry point and one exit point.

This theory is an important contribution to the art of programming because it places necessary restrictions on program structure. The GOTO statement is unacceptable in

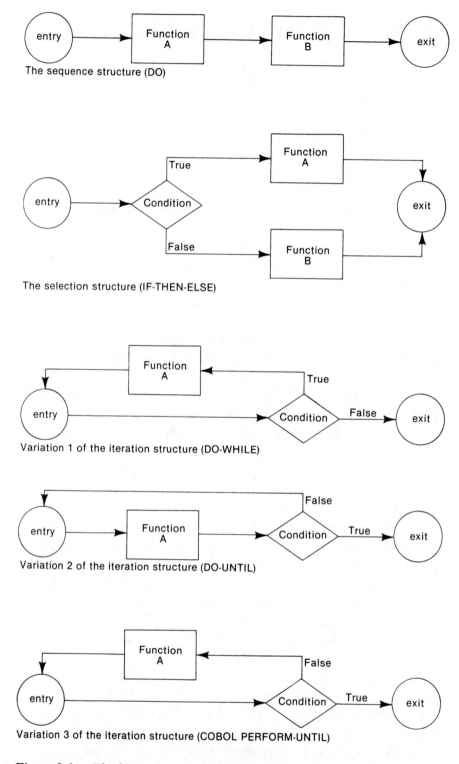

The sequence structure (DO)

The selection structure (IF-THEN-ELSE)

Variation 1 of the iteration structure (DO-WHILE)

Variation 2 of the iteration structure (DO-UNTIL)

Variation 3 of the iteration structure (COBOL PERFORM-UNTIL)

Figure 9-1   The basic structures of structured programming

structured programming. As a result, uncontrolled branching is impossible. This reduces the likelihood of bugs and makes bugs that do occur easier to find and correct. Also, a structured program is easier to read and understand than an unstructured one.

Designing a structured program

The traditional way to design a program is to draw a flowchart for it. However, flowcharts are difficult to create and, once created, difficult to change. It's easy to forget important design details when creating a flowchart. Moreover, the flowchart becomes less and less useful as a program becomes larger and larger.

Fortunately, new methods of program design have been developed as part of the structured programming movement. Although the details vary from one method to another, they all attempt to design a program from the top down. This is referred to as *top-down design*. In addition, these methods provide practical frameworks for implementing programs that take advantage of structured programming theory. As a result, this is also referred to as *structured design*.

Creating a program structure chart   When structured design is used, a *program structure chart* is created. This chart indicates all of the modules required by the program. To create such a chart, the programmer identifies major functional modules within the program. These are divided into their subordinate modules. Division continues until the entire program is divided into functional modules that can be coded with limited difficulty. In contrast, traditional program development methods force you to deal with trivial programming details in the early stages of design.

To create a program structure chart, start with a top-level module that represents the entire program. Then, decide on the one primary functional module that will be performed repeatedly during the execution of the program. Next, think about what functional modules need to be performed before and after the primary module. Then, divide each of these into clearly defined component functions.

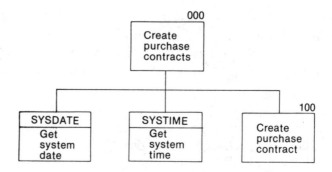

Figure 9-2   The first two levels of the program structure chart for the create-purchase-contracts program (BF1110)

Figure 9-2 shows the top two levels of the program structure chart for the CREATE PURCHASE CONTRACTS program (BF1110) in the brokerage system. The primary functional module that will be performed again and again is CREATE PURCHASE CONTRACT. (Recall from the program overview in figure 8-11 that the program will accept contract data and create new contracts until the operator signals that entry should end.) In addition to the primary module, two others appear at this level. Both represent subprograms. One will be used to get the system date, the other the system time.

Figure 9-3 shows the top three levels of the chart. The two subprograms aren't divided further, but CREATE PURCHASE CONTRACT is. It consists of three subordinate modules: CLEAR WORK FIELDS, GET VERIFIED CONTRACT DATA, and POST CONTRACT DATA.

I continued this process of identifying subordinate modules until I reached the point where I felt confident I could code each module with relative ease. Figure 9-4 is the complete program structure chart for BF1110.

In figure 9-4, notice that I numbered each module with a unique, identifying number. When I code the program, I'll use these numbers to identify sections of code that correspond to each module. Also, I didn't assign numbers to the three subprograms to be used (in addition to SYSDATE and SYSTIME at the top of the chart, I also included a date-edit subprogram, DATEDIT, subordinate to module 600). Finally, you can see that I repeated two modules (931 and 941). In

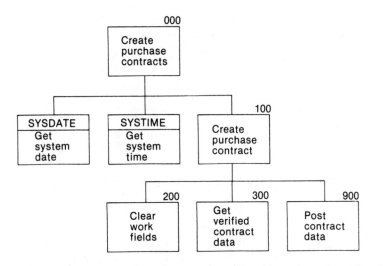

Figure 9-3   The first three levels of the program structure chart for the create-purchase-contracts program

both cases, I shaded the upper right-hand corners of those boxes to indicate they are *shared*, or *common*, modules.

When the programmer is finished, the structure chart shows all of the modules of the program as well as the hierarchical relationships among them. Each module can be coded with one entry point and one exit point using the three structures of structured programming. As a result, this will be a proper program.

I'm sure you immediately recognized the similarity between the program structure chart and the system structure chart I described in chapter 6. In fact, they're developed in a similar manner. The difference is that the system chart represents the modules (programs) of the system, while the program chart represents the modules (routines or paragraphs) of the program. Thus, the entire system is broken down into manageable programming routines.

Creating module documentation   Whether or not you should document the modules of a program before coding them is open for debate. Some people recommend that you do document them using a separate piece of documentation for each module. That supplementary documentation lists the input, output, and processing requirements for each module.

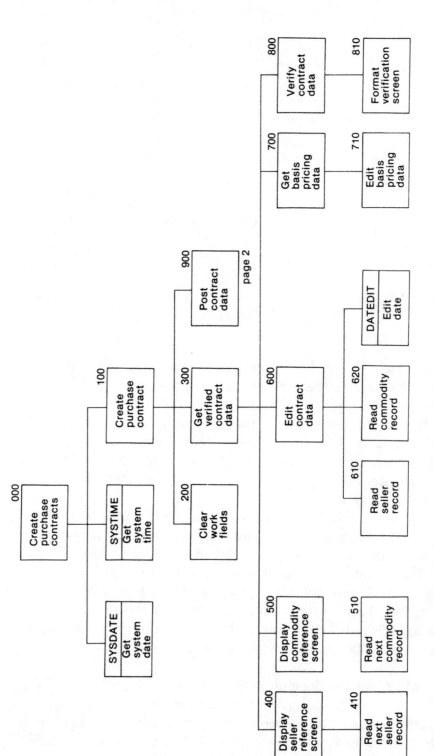

Figure 9-4   The complete program structure chart for the create-purchase-contracts program (part 1 of 2)

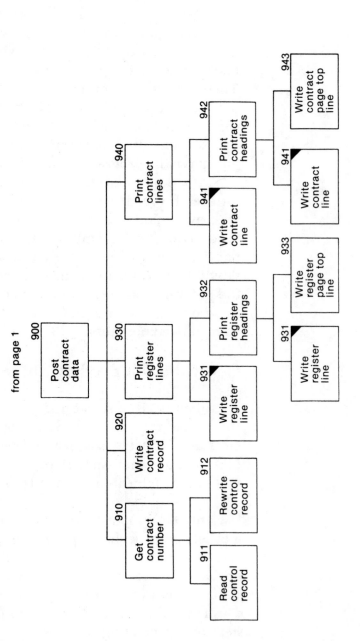

Figure 9-4   The complete program structure chart for the create-purchase-contracts program (part 2 of 2)

Sometimes, the processing is described in the programming language that will be used. Other times, a narrative or pseudocode is used.

In a large shop, you may be required to document the modules of your programs using documents like IPO charts. However, I don't encourage you to do so if you aren't required to. In my experience, I've found that module design documentation of any kind is redundant with the actual program code. So if you develop design documentation for each module, you'll be doing double work. You'll code the module once for documentation and again in the program itself. Then, once the program is complete, the module design documentation is unnecessary. You can throw it away, or you can keep it as part of your program documentation. But if you keep it, you'll have the burden of maintaining it whenever the program needs to be modified.

Now I'm not saying that you shouldn't use pseudocode (or even flowcharts for that matter) to help you clarify in your own mind how a module should work. Do whatever you need to do. But remember that these are your working papers. Once the module is coded and tested, these papers should be thrown away unless they meaningfully supplement the program code. And that won't be often. A clearly coded program that uses the constructs of structured programming is its own documentation.

Conventions for coding a structured program

A primary goal of structured programming is to create code that is as clear and readable as possible. A program that is clear and readable is easier to develop, test, debug, and maintain than one that isn't. Here, then, are some guidelines that will help you create more readable code.

Use only the proper structures A proper program consists of a collection of sequence, selection, and iteration structures arranged with only one entry point and one exit point. But uncontrolled use of GOTOs violates that principle from the start. If you avoid uncontrolled branching and use only the constructs of structured programming, you'll be taking a big step toward creating the most readable and understandable programs you've ever written.

Some programming languages lend themselves neatly to structured coding, but others don't. If you use COBOL and PL/I, for instance, it's easy to create a structured program with no GOTO statements. Both include statements that allow you to code the sequence, selection, and iteration structures directly.

In other programming languages, however, you may have some difficulty coding proper programs. In some versions of BASIC, for example, you have to simulate the structures rather than code them directly because the language doesn't provide the selection or iteration structures. Usually, simulating these structures forces you to use GOTOs. Even if your language doesn't lend itself to structured program development, though, I still encourage you to simulate the proper structures within the limitations of your language. The result will be a longer program, but one that's easier to create and maintain than one that's totally unstructured.

Figure 9-5 illustrates how you can code the top-level control modules of an interactive program without using GOTO statements. This COBOL example is a part of the final source code for program BF1110 of the brokerage system. Whether you know COBOL or not, I hope the clarity of the code in figure 9-5 is apparent. As you can see, this figure only shows the coding for some of the critical control modules of the Procedure Division.

If structured programming is new to you, this partial program may seem obscure. Frankly, the first couple of times I tried my hand at designing and coding a structured COBOL program, I got confused. But that didn't last long. After I designed and coded a few programs, this structure and coding method became second nature to me. Now, I can easily develop the top-level modules for almost any application program very rapidly. So even if you're completely new to the concepts of structured programming, I hope you'll agree that this structured program is a simple and elegant implementation of what could easily become a "rat's nest" using unstructured code.

Use module names that relate to the program structure chart   When you code a module, you should derive the name for the module from the program structure chart. In COBOL, we do this by combining a module's number and name. Thus,

```
 IDENTIFICATION DIVISION.
*
 PROGRAM-ID.       BF1110.
*
 ENVIRONMENT DIVISION.
*
          .
          .
*
 DATA DIVISION.
*
          .
          .
*
 WORKING-STORAGE SECTION.
*
 01  SWITCHES.
*
     05   END-PROGRAM-SW                  PIC X.
          88   END-PROGRAM                          VALUE "Y".
     05   CONTRACT-DATA-VALID-SW          PIC X.
          88   CONTRACT-DATA-VALID                  VALUE "Y".
     05   CONTRACT-VERIFIED-SW            PIC X.
          88   CONTRACT-VERIFIED                    VALUE "Y".
*
          .
          .
*
 PROCEDURE DIVISION.
*
 000-CREATE-PURCHASE-CONTRACTS.
*
     OPEN SHARED PURCHCON
                 SELLERS
                 COMMODTY
                 PCCNTRL
          I-O    SCREEN
          OUTPUT PCONTRCT
                 PCREG.
     CALL "SYSDATE" USING CURRENT-DATE.
     CALL "SYSTIME" USING CURRENT-TIME.
     MOVE "N" TO END-PROGRAM-SW.
     PERFORM 100-CREATE-PURCHASE-CONTRACT
         UNTIL END-PROGRAM.
     CLOSE      PURCHCON
                SELLERS
                COMMODTY
                PCCNTRL
                SCREEN
                PCONTRCT
                PCREG.
     STOP RUN.
```

Figure 9-5   A partial source listing in COBOL for the create-purchase-contracts program
            (part 1 of 3)

```
*
 100-CREATE-PURCHASE-CONTRACT.
*
     PERFORM 200-CLEAR-WORK-FIELDS.
     MOVE "N" TO CONTRACT-DATA-VALID-SW
                CONTRACT-VERIFIED-SW.
     PERFORM 300-GET-VERIFIED-CONTRACT-DATA
         UNTIL CONTRACT-DATA-VALID
             OR END-PROGRAM.
     IF NOT END-PROGRAM
         IF CONTRACT-VERIFIED
             PERFORM 900-POST-CONTRACT-DATA.
*
 200-CLEAR-WORK-FIELDS.
*
 300-GET-VERIFIED-CONTRACT-DATA.
*
     DISPLAY AND READ CONTRACT-SCREEN ON SCREEN
         PFKEYS 5 9 16.
     IF SELECTION = 5
         PERFORM 400-DISPLAY-SELLER
     ELSE
         IF SELECTION = 9
             PERFORM 500-DISPLAY-COMMODITY
         ELSE
             IF SELECTION = 16
                 MOVE "Y" TO END-PROGRAM-SW
             ELSE
                 MOVE "Y" TO CONTRACT-DATA-VALID-SW
                 PERFORM 600-EDIT-CONTRACT-DATA
                 IF      CONTRACT-DATA-VALID
                     AND CONTRACT-PRICE = ZERO
                     PERFORM 700-GET-BASIS-PRICING-DATA.
     IF CONTRACT-DATA-VALID
         PERFORM 800-VERIFY-CONTRACT-DATA.
*
 400-DISPLAY-SELLER-REF-SCREEN.
*
 410-READ-NEXT-SELLER-RECORD.
*
 500-DISPLAY-COMM-REF-SCREEN.
*
 510-READ-NEXT-COMMODITY-RECORD.
*
 600-EDIT-CONTRACT-DATA.
*
 610-READ-SELLER-RECORD.
*
 620-READ-COMMODITY-RECORD.
*
*
*
```

Figure 9-5   A partial source listing in COBOL for the create-purchase-contracts program
          (part 2 of 3)

```
*
*
 700-GET-BASIS-PRICING-DATA.
*
     DISPLAY AND READ BASIS-PRICE-SCREEN ON SCREEN
         PFKEY 16.
     IF SELECTION = 16
         MOVE "N" TO CONTRACT-DATA-VALID-SW
     ELSE
         PERFORM 710-EDIT-BASIS-PRICING-DATA.
*
 710-EDIT-BASIS-PRICING-DATA.
*
 800-VERIFY-CONTRACT-DATA.
*
     PERFORM 810-FORMAT-VERIFICATION-SCREEN.
     DISPLAY AND READ VERIFICATION-SCREEN ON SCREEN
         PFKEYS 1 4 16.
     IF SELECTION = 1
         MOVE "Y" TO CONTRACT-VERIFIED-SW
     ELSE
         IF SELECTION = 4
             MOVE "N" TO CONTRACT-DATA-VALID-SW.
*
 810-FORMAT-VERIFICATION-SCREEN.
*
 900-POST-CONTRACT-DATA.
*
 910-GET-CONTRACT-NUMBER.
*
 911-READ-CONTROL-RECORD.
*
 912-REWRITE-CONTROL-RECORD.
*
 920-WRITE-CONTRACT-RECORD.
*
 930-PRINT-REGISTER-LINES.
*
 931-WRITE-REGISTER-LINE.
*
 932-PRINT-REGISTER-HEADINGS.
*
 933-WRITE-REG-PAGE-TOP-LINE.
*
 940-PRINT-CONTRACT-LINES.
*
 941-WRITE-CONTRACT-LINE.
*
 942-PRINT-CONTRACT-HEADINGS.
*
 943-WRITE-CONT-PAGE-TOP-LINE.
```

Figure 9-5   A partial source listing in COBOL for the create-purchase-contracts program (part 3 of 3)

000-CREATE-PURCHASE-CONTRACTS is the COBOL paragraph name for the top-level module in the program structure chart in figure 9-4. In figure 9-5, you can see that all of the paragraph names in the program are derived in the same way. Also, you can see that they're kept in numeric sequence within the COBOL program.

Two benefits result when you create paragraph names in this way. First, the program structure chart becomes a directory to all of the modules of the resulting program. Then, if you have to maintain a program, you can identify the problem modules by studying the structure chart. Second, because the modules of the program are kept in numeric sequence, you can find the modules in seconds. Compare this with traditional development methods that give you no directory to the names of the program modules and no easy way to find a program module in the source code if you do know its name.

Unfortunately, you can't create program names like 000-CREATE-PURCHASE-CONTRACTS in all programming languages. However, you can approximate this naming technique by using just the module numbers or by using comments instead of actual procedure names to identify modules. We have adapted this naming technique to PL/I, BASIC, FORTRAN, and assembler in the past, so I know that it can be done and that it's worth doing.

Group related data elements   You should group related data elements to give the data section of a program a structure of its own. That way, it will be easier for you to locate data elements in the program when you need to do so. Data elements that you should group include switches, flags, date and time fields, print control fields, subscripts, tables, record descriptions, and print lines.

In figure 9-5, I omitted most of the code that describes the data fields the program uses. Nevertheless, you can see how I grouped all switches together. You'll notice here that I showed the subordination of the three switch fields within the SWITCHES group using indentation, vertical alignment, and level numbers (01 and 05).

Because the same groups of related fields appear again and again in application programs, you should also adopt a standard sequence for them in your programs. For instance, you can code all switches first in the data section of the program, followed by all flags, followed by date and time fields, followed by print control fields, and so on. If all the programmers in your shop use the same sequence of groups, maintaining another programmer's program is a simpler, less frustrating task.

Use descriptive data names   When you create data names, don't abbreviate unnecessarily. Make the names as descriptive as you can within the constraints of your programming language. Also, when data elements are part of the same structure, you should add a prefix or a suffix to each name to identify the group to which the data element belongs. You'll recall from chapter 7 that I use a prefix to identify all data elements that belong to one record description. I can also use suffixes to identify related fields. An example of a suffix is the -SW I added to the names of the switch fields in figure 9-5.

If you work in COBOL, you know that it allows long hyphenated data names, so you can create names that are fully descriptive. But some languages don't have that flexibility. For example, some BASICs restrict you to short data names and don't allow hyphenation. The result is that a program written in a limited BASIC is more confusing and difficult to read than it would be if it were written in COBOL. Unfortunately, there's no solution for this language limitation. You just have to do the best you can with it.

Use comments sparingly   The structure and code of a program should make it easy to understand without comments. As a result, you should use comments infrequently. Comments can be useful if code is obscure, if you're using them to overcome language limitations, or if it's necessary to provide a reference to documentation that's external to the program. But you shouldn't use them to explain something that the code itself should make obvious.

Testing a structured program from the top down

The benefits of structured programming really begin to show themselves as testing begins. First, a structured program is less likely to have bugs to start with than an unstructured program because its modules are independent and have single entry and exit points. Then, if a program does have bugs, they're usually simple clerical errors rather than complex logical errors. So if you approach program development in a sensible way, your testing problems should be minimal. In this topic, I'll describe the most sensible approach to testing that I know of, top-down coding and testing.

The theory of top-down coding and testing   Traditionally, a programmer codes a program in its entirety before he starts testing it. Using this approach, a substantial program is likely to have dozens of bugs in it when testing begins. As a result, testing may turn into a nightmare. It's not unusual, in fact, for the testing phase of program development to take longer than the design and coding phases combined. In contrast, top-down coding and testing can simplify the testing phase dramatically.

When you use top-down coding and testing, you don't code the entire program and then test it. Instead, you code and test in phases. You normally start by coding the top-level module and one or two of its subordinate modules. Then, after you correct any bugs you find, you add one or two more modules and test again. You continue in this way until the entire program has been coded and tested. Because top-down coding and testing always go together, the term *top-down coding* implies *top-down testing*, and vice versa.

To illustrate, figure 9-6 is my top-down development plan for the CREATE PURCHASE CONTRACTS program (BF1110). The module numbers refer to the structure chart in figure 9-4. As you can see, I divided the coding and testing of this program into seven stages. Each is manageable and easy to control. In the first stage, I'll code the top two levels of the program: modules 000 and 100 (the subprograms SYSDATE and SYSTIME are already coded and compiled). In stage 2, I'll add seven modules to the program. In stage 3, I'll add two more. And I'll continue until the entire program is coded and tested. By using top-down coding and testing, I'll shorten the overall development time required for the program.

Using program stubs   When you use top-down testing, you have to use *dummy modules*, or *program stubs*, for the

Stage	Modules Coded	Stub Modules	Description
1	000 100	200 300 900	Code high-level control modules
2	200 300 600 610 620 800 810	400 500 700 900	Code "nominal path" processing modules
3	700 710	400 500 900	Code basis pricing processing modules
4	400 410 500 510	900	Code reference screen display modules
5	900 910 911 912 920	930 940	Code posting control module and file update modules
6	930 931 932 933	940	Code register preparation modules
7	940 941 942 943	None	Code contract preparation modules

Figure 9-6   A top-down testing plan for the create-purchase-contracts program

modules of the program that are called but not yet coded. When I use the test plan in figure 9-6, for example, module 100 in stage 1 calls three subordinate modules, but none of them is done. As a result, each must be coded as a program stub.

A program stub is a module that's coded only for testing purposes. At the very least, a stub should indicate that it was executed. In COBOL, a simple DISPLAY statement is usually

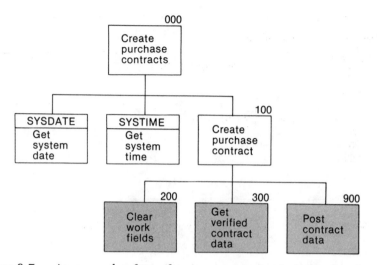

Figure 9-7   An example of top-down program development for the create-purchase-contracts program (part 1 of 6)

adequate for this purpose. In some cases, a stub must also pass control information or data back to the calling module.

Part 1 of figure 9-7 shows the structure of BF1110 during stage one of top-down coding and testing as planned in figure 9-6. The shaded modules are stubs. In this instance, the stubs for modules 200 and 900 do not need to pass any data back to module 100. All they need to do is indicate that they were executed. The stub for module 300, however, must do more.

When the program is finally in production, module 300 will capture all of the data necessary to create a contract and request operator verification. If the data is verified, module 100 should invoke module 900 to post the data. If the data is not verified (in other words, if the operator cancels the entry), module 900 should not be invoked. As a result, module 300 must set a switch that module 100 will evaluate to determine whether or not to call module 900. (In the code in figure 9-5, that switch is CONTRACT-VERIFIED-SW.) If module 100 is to be fully tested, the stub for module 300 must be coded so it will pass both possible switch values back to module 100 in separate entry cycles.

An example of top-down coding and testing   Now that you understand why stubs are necessary and how to use them, I

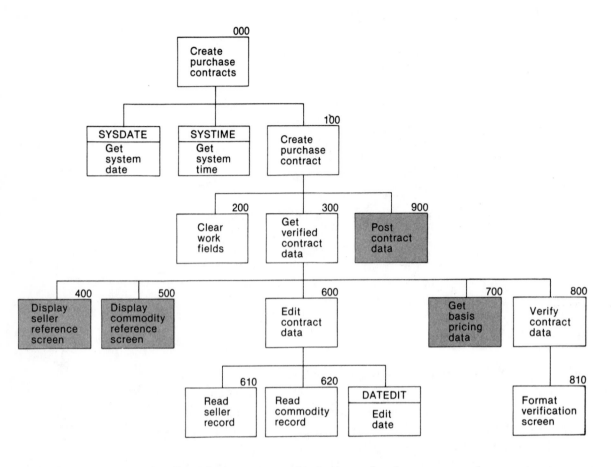

Figure 9-7 An example of top-down program development for the create-purchase-contracts program (part 2 of 6)

can present the complete top-down development sequence for the CREATE PURCHASE CONTRACTS program. After I'm sure that the modules coded in stage 1 work properly, I can move on to stage 2.

Part 2 of figure 9-7 illustrates the second stage of coding and testing. Here, I replaced the stub for module 200 with the actual code of the module. This module consists of simple imperative statements executed in sequence, so it was easy to code. I deferred coding module 900 and kept its stub in my program. Most of my work in stage 2 was coding module 300 and two of its subordinates, modules 600 and 800.

Modules 600 and 800 are on the *nominal path* of the program's processing. The nominal path is the normal execution

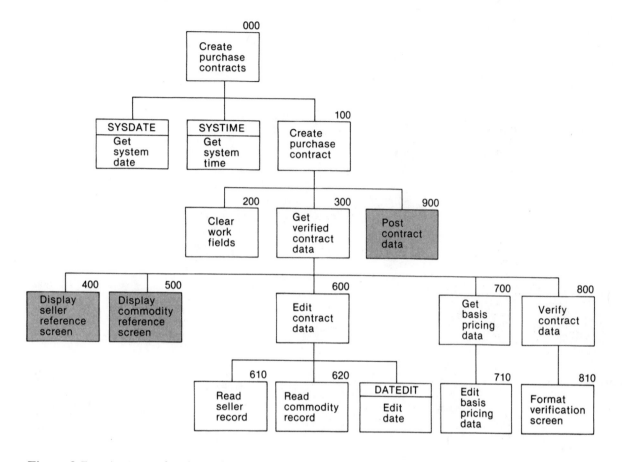

Figure 9-7   An example of top-down program development for the create-purchase-contracts
program (part 3 of 6)

sequence for an entry cycle. In most cases, modules 400, 500,
and 700 won't be entered during contract creation, but
modules 600 and 800 will be done each time a contract is
created. As a result, it made sense to develop them early. To
code module 300, I had to create stubs for modules 400, 500,
and 700. They're shaded in part 2 of figure 9-7.

Part 3 of figure 9-7 illustrates stage 3 of top-down
development. After the modules on the nominal path had
been coded and tested, I moved off the nominal path. I had
decided in my test plan to develop the basis-pricing modules
next. So I replaced the stub for module 700 with its actual
code, added the subordinate module 710 to the program, and
tested both.

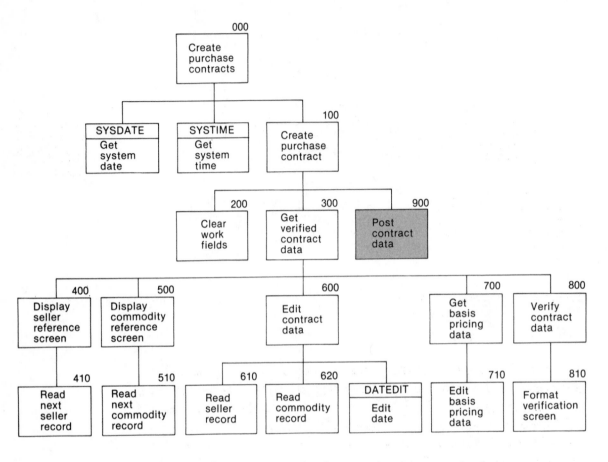

Figure 9-7 An example of top-down program development for the create-purchase-contracts program (part 4 of 6)

Part 4 of figure 9-7 illustrates stage 4. My next step was to develop the reference screen display modules. When I had coded and tested them, all of the program modules subordinate to module 300 were complete.

Next, I turned my attention to posting the data captured and validated by module 300. Part 5 of figure 9-7 shows how I chose to develop the functions subordinate to module 900. In stage 5, I coded module 900 itself and the modules that will update the files the program uses: 910, 911, 912, and 920. To test module 900, I had to add stubs for the register- and contract-preparation modules (930 and 940). After I had tested the control and file-update code in this stage, I turned to the register-preparation module.

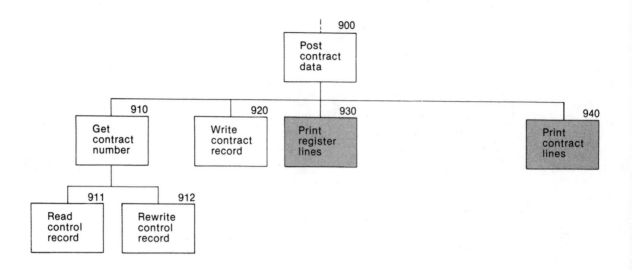

Figure 9-7   An example of top-down program development for the create-purchase-contracts program (part 5 of 6)

Part 6 of figure 9-7 shows the program's structure after I'd coded and tested the register-preparation module and its subordinates in stage 6. All that was left to do then was to replace one stub: contract preparation. After I coded and tested the contract-preparation modules in stage 7, the program was complete. If there were a part 7 to figure 9-7 representing the completely coded program, it would look exactly like figure 9-4, the program structure chart.

The advantages of top-down coding and testing   In contrast to traditional testing, top-down testing proceeds in controlled, manageable steps, as you've just seen. When bugs are discovered, they almost have to be in the modules just added or in the interfaces between them and the old modules.

One major advantage of top-down coding and testing is that it allows you to look for major problems early. As I illustrated, you should first make sure all high-level modules are entered and exited properly and that the interfaces between modules are correct. Next, you can focus on areas of the design you don't feel confident about. You can develop the modules on the program's nominal path early in the development sequence. Then, after you're confident about the major functions of the program, you can worry about details.

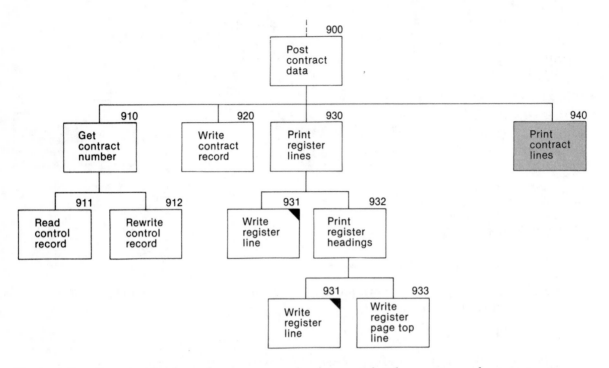

Figure 9-7 An example of top-down program development for the create-purchase-contracts program (part 6 of 6)

By its nature, top-down coding and testing can boost programmer morale. When using traditional methods, coding often drags on for long periods with no feedback to the programmer. Have any omissions been made? Are there any logical deficiencies? Typically, the programmer must wait until the program is completely coded before he can find out. Then, when he does, it can be an ordeal to find and correct the program's bugs. How frustrating. In contrast, top-down testing allows the programmer to get feedback continuously through the coding process and to keep error-correction manageable.

Documenting a structured program

The documentation for a structured program is easy to assemble. It consists of the original program specification used to design the program, the program structure chart, and the program source listing. The original program specification consists of the program overview and any screen or report layouts.

Of course, any specification or design change made during coding and testing should be shown on the related items of documentation. If a new file is to be accessed by a program that isn't on the program's overview, the overview needs to be updated to show it. Similarly, if new program modules are added during coding, they need to be added to the program structure chart.

In practice, modules are often added to programs during coding and testing. As a result, maintaining program structure charts can become burdensome. In our shop, then, we wrote a program that is a structure listing generator. It reads a program's source code and creates a *structure listing* that gives all the information that a system structure chart does. For instance, figure 9-8 shows the structure listing that corresponds to the program structure chart in figure 9-4. Because this technique has saved us hours of tedious chart work, you should consider creating a program like this for your shop.

Discussion

The main objective of structured programming techniques is to improve productivity in all phases of program development. This includes not only initial program development but also on-going program maintenance. If you adopt the design, coding, and testing methods I described in this chapter, I'm confident you'll experience a dramatic improvement in productivity.

To give you some perspective on structured programming and its effects on productivity, let me give you some background. As I see it, development productivity depends on three major factors: (1) your method for system design and analysis, (2) your method for program development, and (3) your computer system. When I developed the brokerage system, I had an optimum set of these factors.

First, I used the analysis and design methods presented in this book. As a result, I was able to complete the design for the system in less than three months including program over-

```
PROGRAM STRUCTURE LISTING:   BF1110                         PAGE   1
                                                            07/22/83

000-CREATE-PURCHASE-CONTRACTS
    SYSDATE
    SYSTIME
    100-CREATE-PURCHASE-CONTRACT
        200-CLEAR-WORK-FIELDS
        300-GET-VERIFIED-CONTRACT-DATA
            400-DISPLAY-SELLER-REF-SCREEN
                410-READ-NEXT-SELLER-RECORD
            500-DISPLAY-COMM-REF-SCREEN
                510-READ-NEXT-COMMODITY-RECORD
            600-EDIT-CONTRACT-DATA
                610-READ-SELLER-RECORD
                620-READ-COMMODITY-RECORD
                DATEDIT
            700-GET-BASIS-PRICING-DATA
                710-EDIT-BASIS-PRICING-DATA
            800-VERIFY-CONTRACT-DATA
                810-FORMAT-VERIFICATION-SCREEN
        900-POST-CONTRACT-DATA
            910-GET-CONTRACT-NUMBER
                911-READ-CONTROL-RECORD
                912-REWRITE-CONTROL-RECORD
            920-WRITE-CONTRACT-RECORD
            930-PRINT-REGISTER-LINES
                931-WRITE-REGISTER-LINE              *
                932-PRINT-REGISTER-HEADINGS
                    931-WRITE-REGISTER-LINE              *
                    933-WRITE-REG-PAGE-TOP-LINE
            940-PRINT-CONTRACT-LINES
                941-WRITE-CONTRACT-LINE              *
                942-PRINT-CONTRACT-HEADINGS
                    941-WRITE-CONTRACT-LINE              *
                    943-WRITE-CONT-PAGE-TOP-LINE
```

Figure 9-8   A structure listing for the create-purchase-contracts program

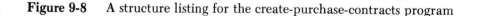

views and related program specifications. Also, the original design was so solid that few changes had to be made to it as I implemented the system.

Second, I used the programming techniques presented in this chapter, so I was able to develop programs at a rapid rate of speed. When I say I developed 128,000 lines of tested COBOL code in six months, I don't mean that I entered every one of the lines into the system. I mean that the final system consisted of that many lines. Rather than enter all of the lines, I used the COPY feature of the COBOL language for all file descriptions, record descriptions, and any other commonly used segments of code including both data descriptions and procedures. Beyond this, I was able to use some programs as the basis for creating others. For instance, the CREATE PURCHASE CONTRACTS program is much like the CREATE SALE CONTRACTS program. As a result, I started the development of the CREATE SALE CONTRACTS program by copying the CREATE PURCHASE CONTRACTS program. In this case alone, I was able to use about 2,000 lines of code taken from the first program. Without reusing segments of code, including large segments of code, there's no way I could have developed the system as rapidly as I did.

Third, I developed the system on a Wang VS computer. In my opinion, this system helped me develop programs at least twice as fast as I could have done on any of the other systems I've worked on, including large mainframes. My workstation provided both word processing and data processing capabilities, so I could modify program overviews, code programs, and test them, all from one device. The system software makes it easy to copy files, modify names within them, and so on, so it was easy for me to make maximum use of tested code taken from other programs. The system points you to your source code when it detects compilation or testing errors so you can correct them right away. And it has many other development features. In short, it has features that you can't duplicate on a smaller system or on a system with a more cumbersome operating system, features that maximize programmer productivity.

On the other hand, the benefits of an efficient analysis and design method and an efficient development computer would be completely lost without an effective way to develop programs. And that's where structured programming comes in.

So if you're interested in more information on the subject, let me recommend two books: *How to Design and Develop COBOL Programs* and the accompanying *The COBOL Programmer's Handbook*, by Paul Noll and Mike Murach. Both present practical techniques for developing structured programs. While their emphasis is on mainframe implementations of batch and CICS applications, they also show you how to develop interactive programs on mini- and microcomputers. (Oddly enough, you can order these books using the order form at the back of this book.)

Yes, other books are available on structured programming techniques. But few give practical ideas for applying the techniques in a specific language. Nevertheless, the more you learn, the better off you are. So get down to your library or bookstore soon.

Terminology

structured programming
sequence structure
selection structure
iteration structure
proper program
top-down design
structured design
program structure chart
shared module
common module
top-down coding
top-down testing
dummy module
program stub
nominal path
structure listing

Objectives

1. Explain the theory of structured programming.

2. In general terms, describe the procedure for designing a structured program.

3. In general terms, describe the coding conventions for structured programs.

4. In general terms, describe the procedure for top-down coding and testing.

Chapter 10

How to document a system

Perhaps the most frustrating phase of a system development project is documenting what you've done. After all, most of us are action people who are excited about getting something accomplished. So doing documentation work, all too often, seems tedious and dull. The job, as far as you're concerned, is done, but you have to keep going over the same old ground.

Unfortunately, documentation is essential. Because businesses change, almost all systems require maintenance as they age. In addition, almost all systems have bugs, some that may not be found until months or years after the installation of the system. When modifications have to be made to a system, it's essential that the persons doing that maintenance have access to complete, accurate, up-to-date documentation.

What I'm going to try to do in this chapter is to give you some general suggestions that will make documentation less of a burden for you. Because the method presented in this book forces you to develop the most important pieces of documentation as you develop the system, documentation shouldn't be as time-consuming as it may have been in the past. And because the pieces of documentation represent an integrated guide to the system, the documentation should be more useful than any you've developed in the past.

In general, when you document a system, you should keep two principles in mind. First, try to keep the documentation to a minimum. To me, the best documentation is just enough to give the reader a full understanding of the workings of a system and no more. If you're confident that your documentation is enough for efficient maintenance of the system, then your documentation is adequate.

Second, try to use documentation that is as easy as possible to create and maintain. If you do this, it's far more likely that the documentation will be done right in the first place and maintained properly thereafter. This means using the output of the analysis, design, and implementation phases as documentation whenever you reasonably can. For instance, the model data flow diagram and the system structure chart, as useful as they are during system development, become the primary pieces of system documentation. Similarly, program specifications are excellent documentation at the program level. Also, the structured program code itself is the best program documentation because it's always up-to-date.

In this chapter, I'm going to show you how to organize your documentation. In addition, I'll suggest some new pieces of documentation that are relatively easy to create and that can make subsequent work on your system an easier task.

In general, I recommend that you divide your system documentation into the three sections shown in figure 10-1. The first will contain general system documentation. The second will contain database documentation. The third will contain program documentation. The easiest way to package these materials is in three-ring binders for 8-1/2 by 11 sheets of paper. And I suggest that you use a separate binder (or set of binders) for each of these three types of documentation. In addition, I suggest that you keep source listings for programs (and sometimes for procedures) in 11 by 14 data binders.

System documentation

The items of system documentation listed in figure 10-1 are important parts of the final documentation for a system. You may want to add others that you feel are critical in your environment, but these items are all that we require. For most systems, a two-inch wide binder is large enough for all of these items.

Section 1:   System documentation

   Model data flow diagram
   System structure chart
   Security concepts and implementation
   Backup concepts and implementation
   User instructions

Section 2:   Database documentation

   File concepts narrative
   Data access diagram
   Data hierarchy diagram
   File and record descriptions in source code
   File-usage cross-reference listing
   File-update cross-reference listing

Section 3:   Program documentation

   Subprogram documentation
      Programmer instructions
      Subprogram listings
      Subprogram-usage cross-reference listing
   Standard segments of source code
   Procedure listings
   Program specifications
      Program overview
      Report layouts
      Screen layouts
   Program implementation documents
      Program structure chart or structure listing
      Program source listing

Figure 10-1   Suggested contents for final system documentation

Model data flow diagram   The model data flow diagram you developed during analysis can be useful to other systems personnel. It reflects your thought processes and your perception of how the system works. As a result, when maintenance work needs to be done, it can give insights into how your system works that might not be available elsewhere.

System structure chart   If you don't include anything else in the system documentation, include the system structure chart. It's this chart that's the blueprint for the physical implementation of the system. It provides the context for the system's functions. It can tell the user at a glance where a function is located and how it fits into the system.

Security concepts and implementation   If you've implemented security measures in your system, your documentation needs to tell what they are and how they work. In particular, if you use a menu structure to control access to parts of the system, this section should clearly and completely explain it. In addition, you should include versions of the system structure chart that reflect that controlled access to the system.

Backup concepts and implementation   As with security measures, your system documentation should treat any special provisions you've made for backup. If the menu structure is used to control backup, you should explain how it works in this section.

User instructions   In my opinion, the fewer external instructions a system requires, the better it is. Nevertheless, some external instructions are usually necessary. These can include a conceptual presentation of how the system works, instructions on hardware operation, and instructions for special conditions such as software or hardware failures, power outages, and so on.

On the other hand, if you've developed an effective system, you shouldn't need to create instructions for normal operating procedures. These should be part of the software itself. When an operator has a question about running a program, the program should answer it, not a manual. Because I developed programs that are largely self-instructional for the brokerage system, the user instructions consist of only 21 pages. In contrast, I've seen user manuals of more than 300 pages for systems that are far less complex than the brokerage system.

Database documentation

If any maintenance work needs to be done on your system, the person doing it must understand your system's database. Using traditional documentation methods, though, database details are often hard to understand. And this can present some serious maintenance problems. As a result, you should make sure that your database documentation is as complete and as clear as possible. For most systems, a two-inch binder is enough for all of the database items that follow.

File concepts narrative   Although I try to avoid narrative whenever I can in system development, it does have its uses. And describing the final implementation of a system's database is one of them. In particular, I find it useful to write a several-page explanation of how and why a system's files are set up the way they are. As you write this, keep in mind that points that may be clear to you after planning and implementing a database can be thoroughly confusing to someone unfamiliar with the system.

Database design diagrams   Just as the model data flow diagram can help others understand your thought processes in organizing the new system, so too can data access and data hierarchy diagrams help others understand why you organized the database as you did. Taken with the file concepts narrative, these diagrams can answer many questions.

File and record descriptions in source code   After you've presented general database design documentation, you need to get down to specifics. For each file or database view, you need to provide the record description in source code. In addition, you should provide any other program-oriented documentation related to the files such as COBOL SELECT statements that describe keying structures. These items should be filed in alphabetical order by file name.

File cross-reference listings   File cross-reference listings are not part of the material you create during system development. However, they are valuable during system maintenance. As a result, I include two types of file cross-reference listings in my final documentation.

The first lists every program that uses a particular file. Logically, I call it the *file-usage cross-reference listing*. The listing is in file sequence. Then, when a file is changed, I can quickly identify every program that may be affected by the change.

The second cross reference I include is a subset of the first. It's the *file-update cross-reference listing*. This listing only shows programs that make changes to a file. It too is in file sequence. If an error is discovered in a file, I use this listing to help find the guilty program.

In practice, you might find it more useful to combine these two cross references into one document. If so, all you have to do is indicate which programs that use a file cause it to be updated.

Program documentation

This section consists of material you created during process specification plus program structure charts and source listings. For a small system, you might package the program documentation in two parts: one three-ring binder for subprogram documentation, program specifications, and structure charts; and a data binder for procedure listings and program source listings. For a large system, though, you might want to package this documentation in several parts. For instance, you might want to keep subprogram documentation in one three-ring binder and program specifications in another binder or set of binders. Then, you might have one data binder for subprogram listings, another binder for procedure listings, and another binder or set of binders for program source listings.

Subprogram documentation For each subprogram in the system, you need to include the programmer instructions that describe what it does and how it's invoked. You should also include the source code of the subprogram. Finally, you might consider a *subprogram-usage cross-reference listing* similar to the file-usage cross reference. This shows which programs use which subprograms. Subprogram documentation should be kept in alphabetical order by subprogram name.

Standard segments of source code When you standardize source code, those segments of code should be part of the system documentation. A typical example is a segment of code that formats a standard report heading. Other segments might be for standard figurative constants, standard data fields (dates, for example), or standard screen descriptions. These items should be kept in alphabetical order by the name of the copy member or file.

Procedure listings   All procedures should be documented here. All that's necessary is a listing of the procedure code and identification of the file that contains it. Procedure listings should be kept in alphanumeric sequence by procedure name. Because procedures are normally only 80 characters wide, they can be kept in an 8-1/2 by 11 binder or in a large data binder.

Program specifications   The final program specifications consist of the documents submitted for program development including any changes made to them during program development. Recall that these specifications consist of the program overview plus report and screen layouts for each program. These specifications should be kept in sequence by program number. Then, the system structure chart will be a guide to the program specifications.

Program implementation documents   For each program in the system, you need to maintain a current copy of the compiled source listing plus a program structure chart or machine-generated structure listing. Because program listings are typically on paper that's too large to include in a three-ring binder, we keep them in data binders. However, we make sure that the listings are kept in the same sequence as their corresponding program specifications.

Because the program structure chart is a guide to the modules of the program, we keep the structure listings in the binders with the program listings. Then, whenever appropriate, a maintenance programmer can work with the data binder alone. In the past, we sometimes kept the structure charts in the binder with the program specifications, but this seemed to separate two pieces of documentation that naturally go together. Before we used structure listings, we taped the structure charts to blank pieces of 11 by 14 paper and kept them in the data binders along with the program listings. This too seemed to be efficient. My recommendation, then, is that you keep the structure chart or listing for a program in the same binder as the source listing for the program.

Discussion

As I said earlier in this chapter, you may include other items as part of your system documentation. For instance, you may require samples of reports or screen images. If you think such items are useful, by all means include them. Nevertheless, the ones I've just listed should make up the bulk of your required documentation.

You can see now that you really aren't creating much documentation when you document the final system. You create a few items, but most of your efforts are toward cleaning up what you already have and organizing it. As a result, documenting a system isn't as tedious a job as it traditionally has been. When I documented the brokerage system, for example, I did so after I installed the system and while I was available for training, enhancements, and debugging. In total, I would say that it took me about two weeks to put the final documentation together, but this was spread over a one- or two-month period.

One thing I like about our documentation system that I haven't found in other systems of documentation is that there's a clear relationship between the system, the database, and the program documentation. In figure 10-2, I've tried to schematically present this relationship. Briefly stated, the model DFD and the system structure chart direct you to the database and program documentation. The program documentation in turn refers to the database documentation and the program implementation documents (the program structure chart and the program source listing). As a result, the documentation of the system is both complete and integrated.

From the more practical point of view of the person doing maintenance, there is a strong relationship between the system structure chart, the program structure chart, and the program source listing. This relationship is shown in figure 10-3. In brief, a module on the system chart points to a module on the program structure chart, which points to a segment of code on the source listing. As a result, the maintenance programmer can identify the program or programs that need to be modified by using the system structure chart and related system documentation. Then, using the module numbers of the system structure chart, he can find the

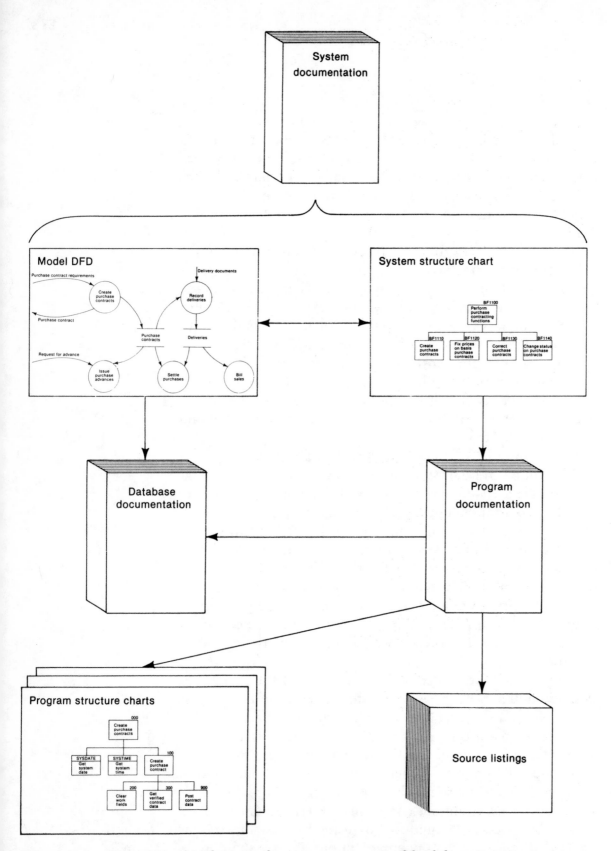

Figure 10-2   The relationships between the major components of final documentation

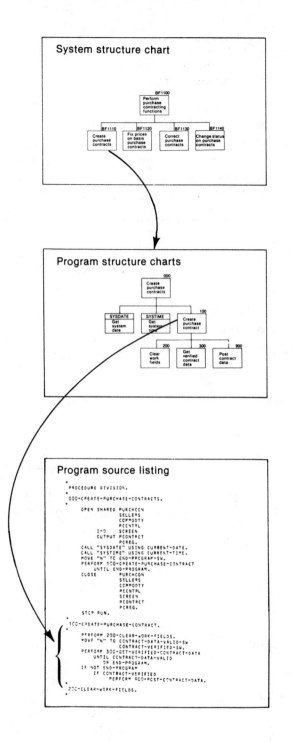

Figure 10-3 The relationships between the structure charts and the source code

related program documentation. Finally, using the program structure chart, he can identify modules that have to be changed in the program, and, using these module numbers, find the related segments of source code. In other words, the final documentation provides a complete directory to every module of source code in the system. As I see it, this is a major benefit of our documentation system.

Terminology

file-usage cross-reference listing
file-update cross-reference listing
subprogram-usage cross-reference listing

Objectives

1. Explain the relationships between system, database, and program documentation.

2. Explain the relationships between the system structure chart, the program structure chart, and the program source code.

Index

Comment Form

Your opinions count

If you have comments, criticisms, or suggestions, I'm eager to get them. Your opinions today will affect our products of tomorrow. If you have questions, you can expect an answer within one week from the time we receive them. And if you discover any errors in this product, typographical or otherwise, please point them out so we can make corrections when the product is reprinted.

Thanks for your help.

Mike Murach
Fresno, California

fold

fold

Book title: How to Design and Develop Business Systems

Dear Mike: _____

_____

_____

_____

_____

_____

_____

_____

_____

fold _____ fold

_____

_____

_____

Name and Title _____

Company (if any) _____

Address _____

City, State, & Zip _____

Fold where indicated and seal.
No postage necessary if mailed in the United States.

fold   fold

NO POSTAGE
NECESSARY
IF MAILED
IN THE
UNITED STATES

BUSINESS REPLY MAIL

First Class   Permit No. 3063   Fresno, CA

POSTAGE WILL BE PAID BY ADDRESSEE

Mike Murach & Associates, Inc.

4222 West Alamos, Suite 101

Fresno, California 93711

fold   fold

Order Form

Our Unlimited Guarantee

To our customers who order directly from us: You must be satisfied. Our books must work for you, or you can send them back for a full refund...no matter how many you buy, no matter how long you've had them.

Name & Title _____

Company (if any) _____

Address _____

City, State, Zip _____

Phone number (including area code) _____

Qty.	Product code and title	Price
CICS and VSAM		
____	CIC1 CICS for the COBOL Programmer: Part 1	$25.00
____	VSAM VSAM for the COBOL Programmer	15.00
COBOL Language Elements		
____	CAG COBOL Advisor's Guide	$100.00
____	SAC1 Structured ANS COBOL: Part 1	20.00
____	SAC2 Structured ANS COBOL: Part 2	20.00
____	RW Report Writer	13.50
COBOL Program Development		
____	DDCP How to Design and Develop COBOL Programs	$20.00
____	CPHB The COBOL Programmer's Handbook	20.00
Other (please specify)		
____	_____	_____
____	_____	_____

Qty.	Product code and title	Price
System Development		
____	DDBS How to Design and Develop Business Systems	$20.00
____	SDCS System Development Case Studies	6.00
____	SDIG System Development Instructor's Guide	35.00
OS Subjects		
____	TSO MVS TSO	$22.50
____	OJCL OS JCL	22.50
____	OSUT OS Utilities	15.00
____	OSDB OS Debugging for the COBOL Programmer	20.00
Assembler Language		
____	ASMD DOS Assembler Language	$22.50
____	ASMO OS Assembler Language	22.50
Other (please specify)		
____	_____	_____
____	_____	_____

☐ Bill me the appropriate price plus shipping and handling (and sales tax in California) for each book ordered.

☐ I want to **save** shipping and handling charges. Here's my check or money order for $_____. California residents, please add 6% sales tax to your total. (Offer valid in the U.S. only.)

☐ Bill the appropriate book prices plus shipping and handling (and sales tax in California) to my
_____ Visa _____ MasterCard:

Card number _____

Valid thru (month/year) _____

Cardowner's signature _____
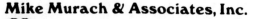
(not valid without signature)

Mike Murach & Associates, Inc.

4222 West Alamos, Suite 101
Fresno, CA 93711 • 209-275-3335

To order more quickly,

Call **toll-free** 1-800-221-5528
In California, call 1-800-221-5527

fold

fold

NO POSTAGE
NECESSARY
IF MAILED
IN THE
UNITED STATES

BUSINESS REPLY MAIL

First Class   Permit No. 3063   Fresno, CA

POSTAGE WILL BE PAID BY ADDRESSEE

Mike Murach & Associates, Inc.

4222 West Alamos, Suite 101

Fresno, California 93711

fold

fold